P9-DVH-549

CliffsStudySolver™

Anatomy & Physiology

By Steven Bassett

Wiley Publishing, Inc.

Published by:
Wiley Publishing, Inc.
111 River Street
Hoboken, NJ 07030-5774
www.wiley.com

Copyright © 2005 Wiley, Hoboken, NJ

Published by Wiley, Hoboken, NJ
Published simultaneously in Canada

Cataloging-in-Publication Data available from the publisher.

ISBN-13: 978-0-7645-7469-6
ISBN-10: 0-7645-7469-8

10 9 8 7 6 5 4 3 2 1

1B/RQ/QT/QV/IN

No part of this publication may be reproduced, stored in a retrieval system, or transmitted in any form or by any means, electronic, mechanical, photocopying, recording, scanning, or otherwise, except as permitted under Sections 107 or 108 of the 1976 United States Copyright Act, without either the prior written permission of the Publisher, or authorization through payment of the appropriate per-copy fee to the Copyright Clearance Center, 222 Rosewood Drive, Danvers, MA 01923, 978-750-8400, fax 978-646-8600, or on the web at www.copyright.com. Requests to the Publisher for permission should be addressed to the Legal Department, Wiley Publishing, Inc., 10475 Crosspoint Blvd., Indianapolis, IN 46256, 317-572-3447, fax 317-572-4355, or online at www.wiley.com/go/permissions.

THE PUBLISHER AND THE AUTHOR MAKE NO REPRESENTATIONS OR WARRANTIES WITH RESPECT TO THE ACCURACY OR COMPLETENESS OF THE CONTENTS OF THIS WORK AND SPECIFICALLY DISCLAIM ALL WARRANTIES, INCLUDING WITHOUT LIMITATION WARRANTIES OF FITNESS FOR A PARTICULAR PURPOSE. NO WARRANTY MAY BE CREATED OR EXTENDED BY SALES OR PROMOTIONAL MATERIALS. THE ADVICE AND STRATEGIES CONTAINED HEREIN MAY NOT BE SUITABLE FOR EVERY SITUATION. THIS WORK IS SOLD WITH THE UNDERSTANDING THAT THE PUBLISHER IS NOT ENGAGED IN RENDERING LEGAL, ACCOUNTING, OR OTHER PROFESSIONAL SERVICES. IF PROFESSIONAL ASSISTANCE IS REQUIRED, THE SERVICES OF A COMPETENT PROFESSIONAL PERSON SHOULD BE SOUGHT. NEITHER THE PUBLISHER NOR THE AUTHOR SHALL BE LIABLE FOR DAMAGES ARISING HEREFROM. THE FACT THAT AN ORGANIZATION OR WEB SITE IS REFERRED TO IN THIS WORK AS A CITATION AND/OR A POTENTIAL SOURCE OF FURTHER INFORMATION DOES NOT MEAN THAT THE AUTHOR OR THE PUBLISHER ENDORSES THE INFORMATION THE ORGANIZATION OR WEB SITE MAY PROVIDE OR RECOMMENDATIONS IT MAY MAKE. FURTHER, READERS SHOULD BE AWARE THAT INTERNET WEBSITES LISTED IN THIS WORK MAY HAVE CHANGED OR DISAPPEARED BETWEEN WHEN THIS WORK WAS WRITTEN AND WHEN IT IS READ.

Trademarks: Wiley, the Wiley Publishing logo, CliffsNotes, the CliffsNotes logo, Cliffs, CliffsAP, CliffsComplete, CliffsQuickReview, CliffsStudySolver, CliffsTestPrep, CliffsNote-a-Day, cliffsnotes.com, and all related trademarks, logos, and trade dress are trademarks or registered trademarks of John Wiley & Sons, Inc. and/or its affiliates. All other trademarks are the property of their respective owners. Wiley Publishing, Inc. is not associated with any product or vendor mentioned in this book.

For general information on our other products and services or to obtain technical support, please contact our Customer Care Department within the U.S. at 800-762-2974, outside the U.S. at 317-572-3993, or fax 317-572-4002.

Wiley also publishes its books in a variety of electronic formats. Some content that appears in print may not be available in electronic books. For more information about Wiley products, please visit our Web site at www.wiley.com.

Note: If you purchased this book without a cover, you should be aware that this book is stolen property. It was reported as "unsold and destroyed" to the publisher, and neither the author nor the publisher has received any payment for this "stripped book."

WILEY is a trademark of Wiley Publishing, Inc.

About the Author

Steven Bassett received his MS from Kearney State College while researching the concentration of a 3-methylhistidine in urine while the patients were subjected to varying conditions. Steven started teaching anatomy and physiology at the high school level in 1978. He has been the lead instructor for anatomy and physiology at Southeast Community College in Lincoln, Nebraska, since 1990. He is currently an adjunct professor in the Physician's Assistance Program at Union College in Lincoln.

Steven has published workbooks, lab manuals, student study guides, and teacher's guides in the field of anatomy and physiology. He is an active member of the Human Anatomy and Physiology Society.

In his spare time, Steven likes to watch his four kids grow while teaching them the value of working harder and studying harder than is expected.

Publisher's Acknowledgments

Editorial

Project Editor: Tere Stouffer

Acquisitions Editor: Greg Tubach

Technical Editor: Dr. Robin Vance

Editorial Assistant: Meagan Burger

Composition

Project Coordinator: Ryan Steffen

Indexer: Steve Rath

Proofreader: Cindy Ballew

Wiley Publishing, Inc. Composition Services

Table of Contents

Pretest

Pretest Questions

The following questions are multiple choice. Determine which choice would be the best answer to the question. The answers to the pretest questions are located at the end of this section.

1. The thumb is considered to be _____ to the little finger.

   A. lateral
   B. medial
   C. anterior
   D. posterior

2. Which of the following terms refers to the lower arm?

   A. Brachium
   B. Antebrachium
   C. Sural
   D. Crural

3. The sural and crural are anatomical terms for the _____.

   A. lower arm
   B. lower leg
   C. upper arm
   D. upper leg

4. The epigastric region is a region that is a part of which of the following?

   A. Thoracic cavity
   B. Abdominopelvic cavity
   C. Pelvic cavity
   D. Cranial cavity

5. The hypogastric region is immediately inferior to the _____ region.

   A. epigastric
   B. umbilical
   C. hypochondriac
   D. iliac

6. The spleen is located in the _____ region.

    A. epigastric

    B. left hypochondriac

    C. right hypochondriac

    D. umbilical

7. The maintenance of an optimal internal environment is termed _____.

    A. positive feedback

    B. negative feedback

    C. homeostasis

    D. metabolism

8. When the body temperature rises, the body has ways to correct the situation and reduce the body temperature back to the normal range. This is an example of _____.

    A. positive feedback mechanism

    B. negative feedback mechanism

    C. homeostasis

    D. catabolism

    E. anabolism

9. Which of the following would be an example of a positive feedback mechanism?

    A. The fluctuation of calcium ions in the bloodstream

    B. The process of childbirth

    C. The rise of insulin production during food intake and the fall of insulin production after food intake

10. In order to determine the atomic mass unit of an atom; you have to do which of the following?

    A. Add the number of protons to the number of electrons

    B. Add the number of protons to the number of neutrons

    C. Add the number of neutrons to the number of electrons

    D. Subtract the number of protons from the number of neutrons

11. Which of the following orbit around the nucleus of an atom?

    A. Electrons

    B. Protons

    C. Neutrons

12. When the number of _____ change, the atom becomes an isotope.

   A. electrons

   B. protons

   C. neutrons

13. Atoms arranged in the first column on the periodic table are arranged in that manner because ____.

   A. they have the same number of protons

   B. they have common characteristics

   C. they have the same number of electrons

   D. they are the smallest atoms

14. Which of the following subatomic particles of an atom cannot change?

   A. Protons

   B. Electrons

   C. Neutrons

15. The last column on the periodic table consists of inert atoms. These atoms will not become _____.

   A. isotopes

   B. ions

   C. stable

   D. radioactive

16. Which of the following represents an ion?

   A. Ca^{2+}

   B. Ca

   C. O_2

   D. ^{16}O

17. Bond the following ions together: Calcium ions and chloride ions.

   A. Ca_2Cl

   B. $CaCl_2$

   C. CaCl

   D. Ca_3Cl_2

18. Which of the following is a polyatomic ion?

   A. N_2

   B. HCO_3^{1-}

   C. CO_2

   D. NaCl

19. Atoms that have the same number of protons but yet a different number of neutrons are called an ____.

    A. ion

    B. isotope

    C. enzyme

    D. anion

20. How many neutrons would ^{23}Na have?

    A. 11

    B. 12

    C. 23

    D. 34

21. The 24 in ^{24}Na represents ____?

    A. the number of protons sodium has

    B. the number of electrons sodium has

    C. the atomic mass unit of sodium

    D. the number of neutrons sodium has

22. Amino acids bonded together will form a ____.

    A. lipid molecule

    B. protein molecule

    C. fat molecule

    D. nucleic acid molecule

23. Molecules that store and process information in the cells are collectively called ____.

    A. nucleic acids

    B. enzymes

    C. proteins

    D. isotopes

24. Carbon dioxide has how many double bonds?

    A. 1

    B. 2

    C. 3

    D. 4

25. A solution that has a pH of 3 is _____ times more acidic than a solution with a pH of 6.

    A. 3
    B. 30
    C. 300
    D. 100
    E. 1,000

26. pH is the measure of _____ in solution.

    A. hydrogen ions
    B. bicarbonate ions
    C. sodium ions
    D. calcium ions

27. A solution that stabilizes the pH of a solution is a _____.

    A. buffer
    B. base
    C. neutral substance

28. Which of the following cell organelles produce cellular energy?

    A. Mitochondria
    B. Ribosomes
    C. Lysosomes
    D. Nucleus

29. Which of the following cell organelles produce protein?

    A. Mitochondria
    B. Ribosomes
    C. Lysosomes
    D. Nucleus

30. Chromosomes are structures that are composed of _____.

    A. molecules of protein
    B. molecules of deoxyribonucleic acid
    C. molecules of enzymes
    D. molecules of lipids

31. The cell membrane is composed of _____ layer/s.

    A. one

    B. two

    C. three

    D. four

32. Which of the following molecules are found only in the outer layer of the cell membrane?

    A. Cholesterol

    B. Protein

    C. Glycolipids

    D. Phospholipids

33. The molecule that transports information from the DNA molecule to the ribosomes of the cell for the purpose of producing protein is _____.

    A. an enzyme

    B. ribonucleic acid

    C. adenosine triphosphate

    D. a lipid

34. The movement of water from an area of high water concentration to an area of low water concentration across a cell membrane is called _____.

    A. a process of diffusion

    B. a process of osmosis

    C. a process of translation

35. If the extracellular fluid consisted of a higher concentration of solutes than the intracellular fluid, the extracellular fluid would be called _____.

    A. hypotonic

    B. hypertonic

    C. isotonic

    D. homeostatic

36. A cell that is exposed to a hypertonic environment will _____.

    A. dehydrate

    B. fill with water

    C. stay in homeostasis

37. The phase of mitosis where the paired chromatids are being pulled apart is called ____.

    A. metaphase
    B. anaphase
    C. telophase
    D. prophase

38. The phase of mitosis where the paired chromatids line up in the center of the nuclear region is called ____.

    A. metaphase
    B. anaphase
    C. telophase
    D. prophase

39. At the end of cell reproduction, one cell will have become ____ cells.

    A. 2
    B. 4
    C. 10
    D. 50

40. Adipose cells belong in which of the following tissue categories?

    A. Epithelial tissue
    B. Muscular tissue
    C. Neural tissue
    D. Connective tissue

41. When viewing cells in the microscope, the cells that have a striped appearance are the ____.

    A. skeletal muscle cells
    B. neural cells
    C. columnar cells
    D. dense cells

42. Cells that contract under involuntary control are ____.

    A. skeletal muscle cells
    B. smooth muscle cells
    C. dense connective cells
    D. glial cells

43. The most superficial layer of the epidermis is the _____.
    A. stratum germinativum
    B. stratum corneum
    C. stratum spinosum
    D. stratum granulosum

44. Melanocytes are cells that _____.
    A. produce a pigment to protect the skin
    B. give the skin its elasticity
    C. protect the skin from dehydration
    D. are involved in hair growth

45. Which layer of skin provides protection against invading pathogens?
    A. Epidermis
    B. Dermis
    C. Hypodermis
    D. Subcutaneous layer

46. Glands that are involved in lubricating the skin are called _____ glands.
    A. sebaceous
    B. merocrine
    C. apocrine
    D. endocrine

47. Glands that produce perspiration for cooling purposes are called _____ glands.
    A. sebaceous
    B. merocrine
    C. apocrine
    D. endocrine

48. A plugged _____ gland may result in pimple formation.
    A. apocrine
    B. merocrine
    C. sebaceous
    D. ceruminous

49. The suture of the skull that articulates the two parietal bones is called the _____ suture.

    A. sagittal
    B. parietal
    C. coronal
    D. temporal

50. How many temporal bones are there in the skull?

    A. 1
    B. 2
    C. 3
    D. 4

51. What is the name of the skull bone that makes up the posterior skull?

    A. Parietal
    B. Occipital
    C. Sphenoid
    D. Ethmoid

52. Which of the following would be considered the "upper" jaw?

    A. Vomer
    B. Maxilla
    C. Sphenoid
    D. Mandible

53. The inferior portion of the nasal septum is called the _____.

    A. perpendicular plate of the ethmoid
    B. vomer
    C. maxilla
    D. zygomatic

54. The "cheek" bones are anatomically known as the _____ bones.

    A. maxillary
    B. mandibular
    C. zygomatic
    D. parietal

55. What is the name of the foramen the spinal cord passes through?

    A. Foramen lacerum

    B. Foramen rotundum

    C. Foramen magnum

    D. Foramen ovale

56. The foramen spinosum is a little posterior to the ____.

    A. foramen ovale

    B. foramen lacerum

    C. jugular foramen

    D. carotid canal

57. The large foramen located on the maxillary bone is the ____.

    A. mental foramen

    B. lacrimal foramen

    C. infraorbital foramen

    D. foramen magnum

58. Vertebra number one that hinges with the skull is called the ____.

    A. axis

    B. atlas

    C. dens

    D. coccyx

59. How many lumbar vertebrae are there?

    A. 5

    B. 7

    C. 12

    D. 24

60. Humans have how many pairs of ribs?

    A. 6

    B. 12

    C. 24

    D. It depends on the sex of the person.

61. Which bone consists of the "elbow"?

    A. Radius

    B. Humerus

    C. Ulna

    D. Scapula

62. How many carpals are there per wrist?

    A. 5

    B. 7

    C. 8

    D. 10

63. The bones that make up the back of the hand are called the _____.

    A. metacarpals

    B. metatarsals

    C. carpals

    D. tarsals

64. Which bone consists of the greater trochanter?

    A. Humerus

    B. Scapula

    C. Femur

    D. Hip

65. Which bone of the lower leg is lateral to the other?

    A. Tibia

    B. Fibula

    C. Femur

    D. Ulna

66. How many tarsal bones are there per ankle?

    A. 5

    B. 7

    C. 8

    D. 9

67. Which muscle is lateral to the palmaris longus?

    A. Flexor carpi radialis
    B. Flexor carpi ulnaris
    C. Biceps brachii
    D. Triceps brachii

68. Which muscle is lateral to the semitendinosus?

    A. Biceps brachii
    B. Biceps femoris
    C. Gracilis
    D. Sartorius

69. What is the name of the major muscle located near the parotid salivary glands and is used for closing the jaw for chewing purposes?

    A. Masseter
    B. Zygomaticus
    C. Platysma
    D. Sternocleidomastoid

70. The main functioning unit of a muscle is the _____.

    A. muscle fiber
    B. muscle cell
    C. muscle sarcomere
    D. muscle sarcolemma

71. During muscle contraction, one of the protein filaments slide. Which of the following is the filament that slides?

    A. Myosin
    B. Troponin
    C. Actin
    D. Tropomyosin

72. Cross-bridges stretch and bond to which of the following?

    A. Myosin
    B. Actin
    C. Troponin
    D. Tropomyosin

73. What is the name of the cells that are responsible for conducting impulses?

    A. Neurons

    B. Glial cells

    C. Schwann cells

    D. Neuroglia

74. Neurotransmitters are released from the ____.

    A. axons

    B. dendrites

    C. soma

75. The most abundant ions located on the outside of the neuron are ____ ions.

    A. calcium

    B. sodium

    C. potassium

    D. chloride

76. Stimulation of the reticular activating system will ____.

    A. put a person to sleep

    B. keep a person alert

    C. cause the brain to cease functioning

    D. cause spinal cord spasms

77. The bundle of nerve fibers that "connect" the left hemisphere with the right hemisphere are called the ____.

    A. corpus callosum

    B. choroid plexus

    C. midbrain

    D. medulla oblongata

78. The thalamus and hypothalamus are part of the ____ region of the brain.

    A. telencephalon

    B. diencephalon

    C. metencephalon

    D. mesencephalon

79. Which of the following is not a part of the nerve plexus system?

    A. Lumbar nerves

    B. Cervical nerves

    C. Thoracic nerves

    D. Brachial nerves

80. The brachial plexus consists of nerves that emerge from the ____.

    A. cervical region
    B. brachial region
    C. cervical and brachial region
    D. cervical and thoracic region

81. The phrenic nerve is part of the ____ plexus.

    A. cervical
    B. brachial
    C. lumbar
    D. sacral

82. The meninges are ____.

    A. specific lobes of the brain
    B. membranes that surround the brain and spinal cord
    C. cavities within the brain
    D. tracts of nerves that ascend and descend the spinal cord

83. Cerebrospinal fluid is produced by the ____.

    A. cerebrum
    B. cerebellum
    C. choroid plexus
    D. corpus callosum

84. Cerebrospinal fluid flows between the ____.

    A. pia mater and arachnoid
    B. arachnoid and dura mater
    C. the brain and pia mater
    D. dura mater and the skull

85. Stimulation of a ____ nerve will cause the pupils of the eyes to constrict.

    A. parasympathetic
    B. sympathetic

86. Stimulation of a ____ nerve will cause the pupils of the eyes to dilate.

    A. parasympathetic
    B. sympathetic

87. Stimulation of the vagus nerve can cause the heart rate to slow down. The vagus nerve is a _____ nerve.

    A. parasympathetic
    B. sympathetic

88. How many pairs of cranial nerves are there?

    A. 6
    B. 10
    C. 12
    D. 24

89. What is the number for the cranial nerve that is involved in sending impulses to the occipital lobe for the interpretation of vision?

    A. I
    B. II
    C. V
    D. VII

90. An affliction of this cranial nerve may result in the condition known as Bell's palsy.

    A. Facial nerve
    B. Vagus nerve
    C. Hypoglossal nerve
    D. Abducens nerve

91. The optic disc of the eye is the region where _____.

    A. the rods and cones of the eye form
    B. light shines through to focus on the retina
    C. blood vessels and the optic nerve emerges from the eye
    D. the lens attaches to the eye itself

92. Which ossicle is connected directly to the ear drum?

    A. Stapes
    B. Malleus
    C. Incus

93. The optic nerve is located _____.

    A. at the center of the back of the eye
    B. a little bit medial to the center of the back of the eye
    C. a little bit lateral to the center of the back of the eye
    D. a little bit inferior to the center of the back of the eye

94. The posterior pituitary gland releases _____.

    A. the antidiuretic hormone
    B. parathormone
    C. calcitonin
    D. adrenocorticotropic hormone

95. Which of the following hormones are involved in sperm production?

    A. Follicle stimulating hormone
    B. Luteinizing hormone
    C. Adrenocorticotropic hormone
    D. Oxytocin

96. The adenohypophysis is known as the _____.

    A. posterior pituitary
    B. anterior pituitary
    C. adrenal cortex
    D. thymus gland

97. Which hormone acts as a negative feedback to calcitonin?

    A. Parathormone
    B. Thyroxine
    C. Adrenalin
    D. Androgens

98. The pancreas releases insulin and also _____.

    A. glucagon
    B. cortisol
    C. androgens
    D. aldosterone

99. What gland produces epinephrine?

    A. Adrenal cortex
    B. Adrenal medulla
    C. Adenohypophysis
    D. Neurohypophysis

100. Erythropoiesis is the process of ____.
   A. the decomposition of red blood cells
   B. the formation of red blood cells
   C. the formation of all the different types of blood cells
   D. the red blood cells delivering oxygen to the body's tissues

101. The breakdown of old red blood cells will eventually form ____.
   A. bilirubin
   B. erythropoietin
   C. bile
   D. fibrinogen

102. A deficiency in ____ will cause the kidney cells to release erythropoietin to begin the process of red blood cell formation.
   A. calcium ions
   B. sodium ions
   C. oxygen
   D. carbon dioxide

103. White blood cells that increase in numbers during an allergic reaction are ____.
   A. monocytes
   B. eosinophils
   C. lymphocytes
   D. neutrophils

104. Which of the following is the most common leukocyte when a patient is in homeostasis?
   A. Neutrophils
   B. Basophils
   C. Eosinophils
   D. Lymphocytes
   E. Monocytes

105. Which leukocyte is typically the first one to respond to a bacterial infection?
   A. Neutrophils
   B. Basophils
   C. Eosinophils
   D. Lymphocytes
   E. Monocytes

106. Heparin is an anticoagulant. It prevents blood from clotting by inhibiting which of the following blood clotting factors?

    A. Calcium ions

    B. Fibrinogen

    C. Platelet thromboplastin factor

    D. Christmas factor

107. Platelets are derived from _____.

    A. thrombocytes

    B. megakaryoblasts

    C. leukocytes

    D. fibrinogen

108. Aspirin is an anticoagulant because it _____.

    A. inhibits vitamin K

    B. inhibits calcium ions

    C. reduces platelet stickiness

    D. inhibits fibrinogen

109. The difference between type A blood cells and type B blood cells is _____.

    A. the blood cell membrane has different glycolipids

    B. the red blood cells have a different shape

    C. the red blood cells metabolize material differently

    D. all of the above

110. When discussing blood, the agglutinin is the specific term for the _____ found in the plasma of blood.

    A. glycolipids

    B. glycoproteins

    C. antibodies

    D. antigens

111. What is the term for the glycolipid on the surface of red blood cells?

    A. Agglutinogen

    B. Agglutinin

    C. Antibody

    D. Rh factor

112. Packed blood consists of ____.

    A. red blood cells only

    B. red blood cells and white blood cells

    C. red blood cells, white blood cells, and plasma

113. Can type A plasma be donated to a type AB patient?

    A. Yes

    B. No

114. A person with type A blood cannot donate to a person with type B blood. This is because the ____ of the type B recipient will be activated by the ____ of the donor. When this happens, blood will clump.

    A. agglutinogens; agglutinins

    B. agglutinins; agglutinogens

    C. glycolipids; antigens

    D. antigens; glycolipids

115. Blood on the left side of the heart is ____.

    A. deoxygenated blood

    B. on its way to the lungs

    C. oxygenated blood

116. The right ventricle of the heart pumps blood ____.

    A. to the lungs

    B. to the body

    C. to the left ventricle

    D. to the right atrium

117. Cardiac cells are found in the ____ layer of the heart.

    A. epicardial

    B. myocardial

    C. endocardial

118. The sinoatrial node (pacemaker) is located in the ____.

    A. left atrium

    B. left ventricle

    C. right atrium

    D. right ventricle

119. The QRS complex of an ECG represents the depolarization of the ____.

    A. atria

    B. ventricles

    C. both atria and ventricles

120. Stimulation of the ____ will result in ventricular contraction.

    A. bundle branches

    B. Purkinje fibers

    C. sino-atrial node

    D. atrio-ventricular node

121. Blockage of the thoracic duct will hinder the flow of ____.

    A. lymph to the left subclavian vein

    B. lymph back to the heart

    C. lymph to the lower extremities of the body

    D. lymph to the abdominal organs

122. Lymphocytes are located in all of the following except the ____.

    A. tonsils

    B. spleen

    C. thymus

    D. brain

123. Which of the following primarily produces antibodies?

    A. T cells

    B. B cells

    C. NK cells

    D. Neutrophils

124. T cells and B cells are a type of ____.

    A. monocyte

    B. lymphocyte

    C. neutrophil

    D. macrophage

125. When the thymus gland fails to produce thymosin, there will be a decreased number of ____.

    A. monocytes

    B. T cells

    C. lymphocytes

    D. B cells

126. Blocking T cell activity would cause a decrease in ____.

    A. antibody production by the B cells

    B. viral activity

    C. monocyte activity

    D. the active immunity process

127. Air enters into the trachea by passing through the opening to the trachea called the ____.

    A. oropharynx

    B. laryngopharynx

    C. glottis

    D. cricoid

128. The palatine tonsils are located ____.

    A. in the nasopharynx region

    B. in the oropharynx region

    D. in the laryngopharynx region

129. The respiratory tubes that branch off the trachea are called the ____.

    A. bronchioles

    B. primary bronchi

    C. carina

130. Blood capillaries that surround the ____ will absorb oxygen.

    A. alveoli

    B. lungs

    C. villi

    D. lobules

131. The diaphragm muscle moves ____ in order for inhalation to occur.

    A. downward

    B. upward

132. In order to inhale, the ____.

    A. thoracic cavity must decrease in size thus increasing thoracic pressure

    B. thoracic cavity must increase in size thus decreasing thoracic pressure

    C. thoracic cavity must decrease in size thus decreasing thoracic pressure

    D. thoracic cavity must increase in size thus increasing thoracic pressure

133. What percentage of the carbon dioxide produced will be exhaled?

    A. 7%

    B. 23%

    C. 70%

    D. 90%

134. Which of the following act as a buffer inside the red blood cell?

    A. Hemoglobin

    B. Chloride ion

    C. Sodium ion

    D. Iron

135. An increase in carbon dioxide will result in _____ in hydrogen ions.

    A. an increase

    B. a decrease

136. What is the name of the salivary glands that are located near the masseter muscle?

    A. Sublingual

    B. Submandibular

    C. Parotid

137. The _____ closes over the opening of the trachea to prevent food from going down the trachea.

    A. uvula

    B. epiglottis

    C. esophageal sphincter

    D. laryngeal sphincter

138. The anatomical name for the canine teeth is _____.

    A. cuspid

    B. bicuspid

    C. premolar

    D. molar

139. Which of the following is the correct sequence for food passing through the small intestine?

    A. Duodenum, jejunum, ileum

    B. Jejunum, duodenum, ileum

    C. Jejunum, ileum, duodenum

    D. Duodenum, ileum, jejunum

140. The structures inside the small intestine that are involved in absorbing nutrients into the bloodstream are called _____.

    A. alveoli
    B. villi
    C. Kuppfer cells
    D. lobules

141. Most of the digestion occurs in the _____.

    A. mouth
    B. stomach
    C. small intestine
    D. large intestine

142. Which of the following is an accessory structure of digestion?

    A. Pancreas
    B. Spleen
    C. Colon
    D. Small intestine

143. What is the function of bile?

    A. Bile will digest fat.
    B. Bile will emulsify fat.
    C. Bile will absorb fat.
    D. Bile is a waste product and serves no function.

144. What organ stores bile?

    A. Gallbladder
    B. Liver
    C. Pancreas
    D. Stomach

145. The small intestine joins the large intestine at the _____.

    A. cecum
    B. transverse colon
    C. descending colon
    D. sigmoid

146. The function of the large intestine is to ____.

    A. get rid of waste

    B. house bacteria, which produce vitamin K

    C. reabsorb water

    D. all of the above

147. The appendix is attached to the ____.

    A. cecum

    B. ascending colon

    C. sigmoid colon

    D. descending colon

148. The hormone, ____ causes the liver to produce bile and the hormone, ____ causes the gallbladder to contract, thereby releasing bile into the small intestine.

    A. cholecystokinin; secretin

    B. secretin; cholecystokinin

    C. bilirubin; lipase

    D. lipase; bilirubin

149. Which of the following hormones is involved in getting the pancreas to release buffers into the small intestine?

    A. Cholecystokinin

    B. Secretin

    C. Glucagon

    D. Insulin

150. Which of the following hormones causes the stomach to produce acid?

    A. Cholecystokinin

    B. Secretin

    C. Gastrin

    D. Glucagon

151. Glycolysis is a metabolic process that occurs in the ____.

    A. mitochondria

    B. cytosol of the cell

    C. nucleus

    D. ribosomes

152. The Krebs reactions are a series of metabolic reactions that occur in the ____.

A. mitochondria

B. cytosol of the cell

C. nucleus

D. ribosomes

153. Which of the following molecules transport hydrogen ions to the electron transport system for the purpose of buffering and ATP production?

A. Cholesterol

B. HDL

C. Glucose

D. NADH

154. Which of the following is the vitamin that becomes NAD?

A. Niacin

B. Thiamine

C. Folic acid

D. Cobalamin

155. The absorption of iron for the purpose of producing hemoglobin requires the presence of ____.

A. vitamin C

B. vitamin D

C. niacin

D. vitamin B_{12}

156. Which vitamin is necessary for the blood clotting process?

A. A

B. D

C. C

D. K

157. The cholesterol that is transported to the liver to be incorporated into bile is known as ____ cholesterol and is transported by ____.

A. good; LDL

B. good; HDL

C. bad; LDL

D. bad; HDL

158. LDL molecules have a tendency to drop cholesterol off in the ____.

    A. liver
    B. gallbladder
    C. arteries
    D. kidneys

159. Good cholesterol has a different molecular structure than bad cholesterol.

    A. True
    B. False

160. Which of the following urinary tubes exit the kidneys?

    A. Urethra
    B. Ureter
    C. Pelvic tubes
    D. Conical tubes

161. Which of the following urinary tubes exit the urinary bladder?

    A. Urethra
    B. Ureter
    C. Pelvic tubes
    D. Conical tubes

162. Which kidney sits higher in the body than the other?

    A. Left
    B. Right

163. The first part of the nephron is the ____.

    A. proximal convoluted tubule
    B. nephron loop
    C. glomerular capsule
    D. renal pyramid

164. The kidneys are made of the cortex region and the medulla region. Most of the nephrons are in the ____ region of the kidneys.

    A. cortex
    B. medulla

165. Waste products are forced out of the glomerular capillaries into the ____ of the nephron.
     A. proximal convoluted tubule
     B. distal convoluted tubule
     C. nephron loop
     D. glomerular capsule

166. One function of the kidneys is to "cleanse" the blood or filter it. This process occurs at the ____.
     A. proximal convoluted tubule and vasa recta
     B. glomerular capillaries and glomerular capsule
     C. nephron loop and vasa recta
     D. renal artery and renal vein

167. Approximately what percentage of the water that enters the kidneys is put back into the bloodstream to prevent dehydration?
     A. 1%
     B. 10%
     C. 50%
     D. 99%

168. After filtration, which vessels consists of "cleaner" blood?
     A. Renal artery
     B. Renal vein

169. What is the name of the tube that transports sperm cells from the testes to the penile urethra?
     A. Ductus deferens
     B. Rete testis
     C. Epididymis
     D. Seminiferous tubules

170. The first gland the sperm cells swim past is the ____.
     A. prostate
     B. seminal vesicle
     C. bulbourethral

171. A hormone from the pituitary gland initiates the production of sperm cells. What is the name of this hormone?

   A. Luteinizing hormone

   B. Follicle stimulating hormone

   C. Gonadotropin hormone

   D. Gametogenesis hormone

172. Successful fertilization of the egg occurs _____.

   A. in the uterus

   B. in the ovaries

   C. in the distal two-thirds of the uterine tubes

   D. in the vagina

173. When a follicle ruptures, it releases the egg. This ruptured follicle is now called a _____.

   A. corpus callosum

   B. corpus luteum

   C. corpus albicans

   D. corpus cavernosum

174. The ruptured follicle will secrete _____.

   A. progesterone

   B. follicle stimulating hormone

   C. luteinizing hormone

175. A decrease in _____ will initiate the menstrual cycle.

   A. progesterone

   B. follicle stimulating hormone

   C. luteinizing hormone

176. A decrease in _____ will initiate menopause.

   A. progesterone

   B. follicle stimulating hormone

   C. luteinizing hormone

177. What releases the human chorionic gonadotropin hormone?

   A. The zygote

   B. The placenta

   C. The uterus

   D. The zygote and placenta

   E. All of the above

Key to Pretest Questions

1. A
2. B
3. B

If you missed numbers 1 through 3, refer to page 42 (Superficial Landmarks).

4. B
5. B
6. B

If you missed numbers 4 through 6, refer to page 51 (Abdominopelvic Regions).

7. C
8. B
9. B

If you missed numbers 7 through 9, refer to pages 58–59 (Homeostasis, Negative Feedback Mechanism, Positive Feedback Mechanism).

10. B
11. A
12. C

If you missed numbers 10 through 12, refer to page 65 (Atoms).

13. B
14. A
15. B

If you missed numbers 13 through 15, refer to page 66 (The Periodic Table).

16. A
17. B
18. B

If you missed numbers 16 through 18, refer to pages 67–69 (Ions and Polyatomic Ions).

19. B
20. B
21. C

If you missed numbers 19 through 21, refer to page 71 (Isotopes).

22. B
23. A
24. B

If you missed numbers 22 through 24, refer to page 74 (Organic Molecules).

25. E
26. A
27. A

If you missed numbers 25 through 27, refer to pages 78–79 (pH Concepts, Buffers).

28. A
29. B
30. B

If you missed numbers 28 through 30, refer to page 83 (Cell Organelles).

31. B
32. C
33. B

If you missed numbers 31 through 33, refer to page 84 (Cell Membrane).

34. B
35. B
36. A

If you missed numbers 34 through 36, refer to page 85 (Osmosis).

37. B
38. A
39. A

If you missed numbers 37 through 39, refer to page 88 (Cell Reproduction).

40. D
41. A
42. B

If you missed numbers 40 through 42, refer to page 95 (The Four Tissue Groups).

43. B
44. A
45. A

If you missed numbers 43 through 45, refer to page 109 (The Skin).

46. A
47. B
48. C

If you missed numbers 46 through 48, refer to pages 110–111 (Nails, Glands).

49. A
50. B
51. B

If you missed numbers 49 through 51, refer to page 115 (Bones of the Cranium).

52. B
53. B
54. C

If you missed numbers 52 through 54, refer to page 117 (Bones of the Face).

55. C
56. A
57. C

If you missed numbers 55 through 57, refer to page 120 (Foramen of the Skull).

58. B
59. A
60. B

If you missed numbers 58 through 59, refer to page 122 (The Thoracic Cage).

61. C
62. C
63. A

If you missed numbers 61 through 63, refer to page 129 (The Pectoral Girdle and Upper Limbs).

64. C
65. B
66. B

If you missed numbers 64 through 66, refer to page 133 (The Pelvic Girdle and Lower Limbs).

67. A
68. B
69. A

If you missed numbers 67 through 69, refer to pages 141–147 (Select Muscles).

70. C
71. C
72. B

If you missed numbers 70 through 72, refer to page 150 (Muscle Structure).

73. A
74. A
75. B

If you missed numbers 73 through 75, refer to page 156 (The Impulse).

76. B
77. A
78. B

If you missed numbers 76 through 78, refer to page 167 (The Brain).

79. C
80. C
81. A

If you missed numbers 79 through 81, refer to page 176 (Nerve Plexuses).

82. B
83. C
84. A

If you missed numbers 82 through 84, refer to page 179 (Protecting the Central Nervous System).

85. A
86. B
87. A

If you missed numbers 85 through 87, refer to page 186 (The Spinal Nerves).

88. C
89. B
90. A

If you missed numbers 88 through 90, refer to page 187 (The Cranial Nerves).

91. C
92. B
93. B

If you missed numbers 91 through 93, refer to page 195 (Introducing the Five Senses).

94. A
95. A
96. B

If you missed numbers 94 through 96, refer to page 207 (The Pituitary Gland).

97. A

98. A

99. B

If you missed numbers 97 through 99, refer to page 209 (Other Endocrine Glands).

100. B

101. A

102. C

If you missed numbers 100 through 102, refer to page 218 (Red Blood Cells).

103. B

104. A

105. A

If you missed numbers 103 through 105, refer to page 219 (White Blood Cells).

106. A

107. B

108. C

If you missed numbers 106 through 108, refer to page 221 (Platelets and Platelet Response).

109. A

110. C

111. A

If you missed numbers 109 and 111, refer to page 224 (Glycolipids and Blood Typing).

112. A

113. B

114. B

If you missed numbers 112 through 114, refer to pages 224–230 (Donating Packed Cells, Donating Whole Blood, Donating Plasma).

115. C

116. A

117. B

If you missed numbers 115 through 117, refer to page 239 (Structure of the Internal Heart).

118. C

119. B

120. B

If you missed numbers 118 through 120, refer to page 245 (The Electrocardiogram [ECG]).

121. A
122. D
123. B

If you missed numbers 121 through 123, refer to page 265 (Lymph and Lymph Vessels).

124. B
125. B
126. A

If you missed numbers 124 through 126, refer to page 266 (The Lymphatic System and Defense).

127. C
128. B
129. B

If you missed numbers 127 through 129, refer to page 273 (The Respiratory Organs).

130. A
131. A
132. B

If you missed numbers 130 through 132, refer to page 275 (The Process of Inhaling and Exhaling).

133. B
134. A
135. A

If you missed numbers 133 through 135, refer to page 277 (The Chloride Shift).

136. C
137. B
138. A

If you missed numbers 136 through 138, refer to page 287 (The Mouth).

139. A
140. B
141. C

If you missed numbers 139 through 141, refer to page 292 (The Small Intestine).

142. A
143. B
144. A

If you missed numbers 142 through 144, refer to pages 296–297 (The Pancreas, The Liver and Gallbladder).

145. A
146. D
147. A

If you missed numbers 145 through 147, refer to page 300 (The Large Intestine).

148. B
149. B
150. C

If you missed numbers 148 through 150, refer to page 303 (Hormones of the Digestive System).

151. B
152. A
153. D

If you missed numbers 151 through 153, refer to page 308 (Metabolism of Carbohydrates).

154. A
155. D
156. D

If you missed numbers 154 through 156, refer to page 315 (Vitamins and Metabolism).

157. B
158. C
159. B

If you missed numbers 157 through 159, refer to page 317 (Cholesterol and Metabolism).

160. B
161. A
162. A

If you missed numbers 160 through 162, refer to page 321 (The Functions and Structures of the Urinary System).

163. C
164. A
165. D

If you missed numbers 163 through 165, refer to page 323 (The Internal Structures of the Kidneys).

166. B
167. D
168. B

If you missed numbers 166 through 168, refer to page 327 (Function of the Nephron).

169. A

170. B

171. B

If you missed numbers 169 through 171, refer to pages 335–336 (The Male Reproductive System and Sperm Cells, The Three Glands of the Male Reproductive System).

172. C

173. B

174. A

If you missed numbers 172 through 174, refer to page 339 (The Female Reproductive System and Egg Cells).

175. A

176. B

177. D

If you missed numbers 175 through 177, refer to page 341 (Hormones Associated with the Female Reproductive System).

Chapter 1

Introductory Anatomical Terminology and Physiological Concepts

Most of the terms used in anatomy and physiology are of either Greek or Latin origin. (In fact, the term *anatomy* is derived from a Greek word that means to "cut open.") These terms are descriptive, and while they may appear to be rather difficult in the beginning, they soon will be second nature to you. You'll find yourself speaking the language of anatomy and physiology in no time.

Directional Terms

Have you ever been in a situation where someone is giving you directions and says something like this, "When you come to corner, turn right"? When you're coming from one direction, turning right may be north. Coming from the opposite direction, turning right may be south. It would be far better to receive directions in this manner, "When you come to the corner, turn north." This way, no matter which direction you're traveling, north is always north.

The same issue applies when describing various aspects of the body. The following is a list of terms that are commonly used when discussing the body, and these are accurate terms regardless how the body is positioned or how you are looking at the body.

❏ **Superior** means moving from one point to another going toward the head. Do not use the word "up." Here is an example: the patient's nose is superior to their mouth. To get to the nose from the mouth, you have to move in a head-like manner—even if the patient happens to be standing on his or her head.

❏ **Inferior** means moving from one point to another going toward the feet. Do not use the word "down." Here is an example: The patient's chin is inferior to their mouth. To get to the chin from the mouth, you have to move toward the feet—even if the patient happens to be standing on his or her head.

❏ **Medial** means moving from one point to another going toward the midline of the body. Do not use the word "inside." Here is an example: the patient's big toe is medial to the little toe. The big toe is not on the "inside" of the foot. To get to the inside of the foot, one would have to make an incision and actually cut into the foot.

❏ **Lateral** means moving from one point to another going away from the midline of the body. Do not use the word "outside." Here is an example: the patient's little toe is lateral to the big toe.

❏ **Anterior** means the point of reference you are referring to is located on the front side of the body. For example: your chest is anterior to your upper back.

❏ **Posterior** means the point of reference you are referring to is located on the back side of the body. For example: your gluteus maximus is on the posterior side of the body.

Example Problems

Fill in the following blanks using one of the previously discussed directional.

1. The knee is _____ to the hip bones.

    answer: inferior

2. The ears are _____ to the nose.

    answer: lateral

3. The shoulder is _____ to the elbow.

    answer: superior

4. The lips are _____ to the nose.

    answer: inferior

5. The elbow is on the _____ side of the arm.

    answer: posterior

Anatomical Position

The **anatomical position** for the human body is when the patient is standing with the palms of their hands facing anterior. This position is necessary for ease of study because it is in this position that the two bones that make up your lower arm are parallel to each other. In this manner, your thumbs will be lateral to your little finger. Regardless of the position the patient is in, always think in terms of anatomical position. Even if a patient is standing in front of you with his arms crossed, you still view his thumb as being lateral to his little finger.

Example Problems

Fill in the following blanks (using one of the previously discussed directional terms) while referring to Figure 1-1.

1. Point 1 is _____ to point 2.

    answer: superior

2. Point 3 is _____ to point 4.

    answer: lateral

3. Point 5 is _____ to point 6.

    answer: lateral

4. Point 7 is _____ to point 8.

    answer: anterior

5. Point 9 is _____ to point 10.

    answer: medial

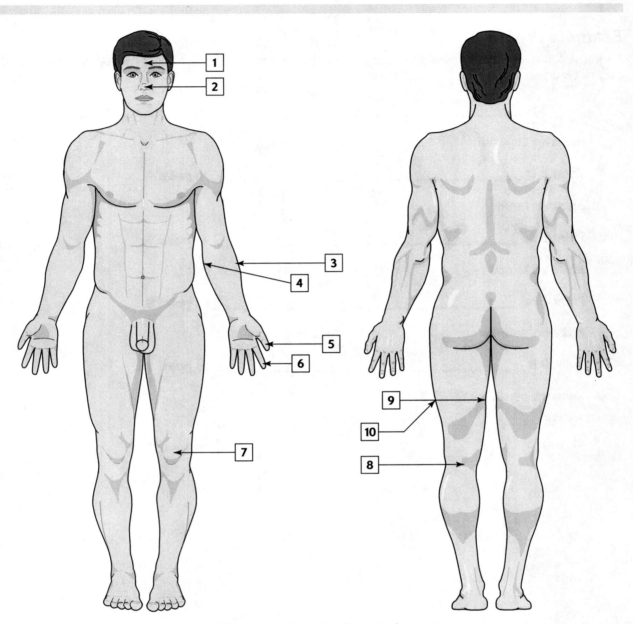

Figure 1-1: Directional terminology.

Work Problems

Fill in the blanks using one of the six directional terms previously discussed.

1. The bulgy part of your elbow is located on the _____ side of the arm.

2. Your fingernails are located on the _____ side of the finger.

3.   Your kneecap is located on the _____ side of the body.

4.   Your naval is located on the _____ side of the body.

5.   Your gluteal region is located on the _____ side of the body.

6.   Your heel bone is located on the _____ side of the foot.

7.   Your eyes are located _____ and _____ to the tip of your nose.

8.   Your naval is located _____ and _____ to the right nipple region.

9.   Your ears are _____ to your nose.

10. Regardless of the position of the body, the thumbs (digit 1) are always _____ to digit 2.

Worked Solutions

1.   **posterior**

2.   **posterior**

3.   **anterior**

4.   **anterior**

5.   **posterior**

6.   **posterior**

7.   **lateral and superior (superior and lateral)**

8.   **inferior and medial (medial and inferior)**

9.   **lateral**

10. **lateral**

Superficial Landmarks

When two people are talking about cars, they typically use car terminology. Both people understand each other as long as both understand car terminology. When two people are talking about computers, they typically use computer terminology. Both people understand each other as long as both understand computer terminology.

The discussion of the human body isn't any different than the discussion of other topics. You just need to know and understand the language. The language used in science is of Latin or Greek origin.

Table 1-1 lists many terms associated with the superficial regions of the body. The column labeled "term" is the scientific term and the "location" is described in laymen's terms.

Table 1-1 Superficial Landmarks			
Term	**Location**	**Term**	**Location**
Frons	Forehead	Brachium	Upper arm
Otic	Ear	Antecubital	Anterior elbow
Oris	Mouth	Cubital (olecranon)	Posterior elbow
Oculus	Eye	Antebrachium	Lower arm
Mentis	Chin	Carpal	Wrist
Nasus	Nose	Pollex	Thumb
Bucca	Cheek	Femur	Thigh
Occipital	Back of head	Patella	Anterior knee
Cervical	Neck	Popliteal	Posterior knee
Axilla	Armpit	Crural (crus)	Anterior lower leg
Thoracic	Chest	Sural (sura)	Posterior lower leg
Abdomen	Abdomen	Calcaneus	Heel region
Umbilical	Naval	Tarsal	Ankle
Costal (dorsum)	Upper back	Hallux	Big toe
Lumbar	Lower back	Inguinal	Groin region
Gluteal cleft	Crease between the left and right buttocks	Gluteal fold	Fold between the buttocks and upper thigh

Example Problems

Look at Table 1-1 and review the terms. Then use those terms to identify the numbered areas associated with Figures 1-2 through 1-5.

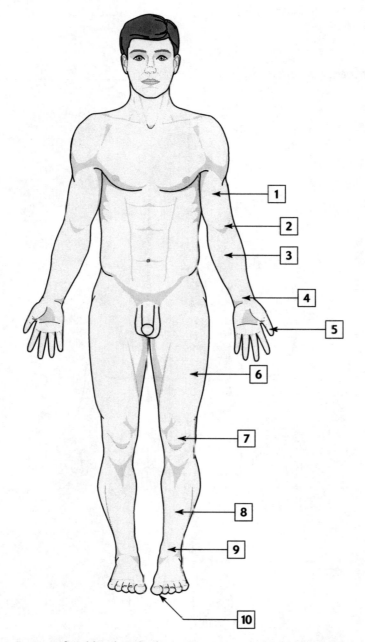

Figure 1-2: Superficial landmarks (anterior view; arm and leg superficial terms).

1. Body area number 1 is called _____.

    answer: brachium

2. Body area number 3 is called _____.

    answer: antebrachium

3. Body area number 5 is called _____.

    answer: pollex

4. Body area number 7 is called _____.

    answer: patella

5. Body area number 9 is called _____.

    answer: tarsal

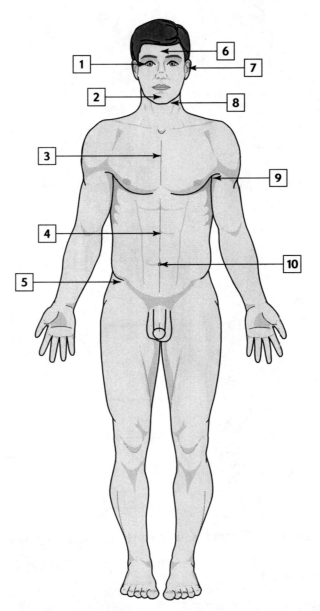

Figure 1-3: Superficial landmarks (anterior view; head and torso superficial terms).

6. Body area number 2 is called _____.

    answer: mentis

7. Body area number 4 is called _____.

    answer: abdomen

8. Body area number 6 is called _____.

    answer: frons (frontal)

9. Body area number 8 is called _____.

    answer: cervical

10. Body area number 10 is called _____.

    answer: umbilical

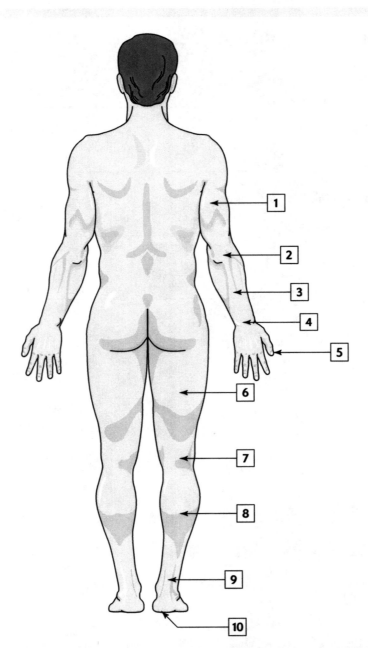

Figure 1-4: Superficial landmarks (posterior view; arm and leg superficial terms).

11. Body area number 1 is called _____.

    answer: brachium

12. Body area number 3 is called _____.

    answer: antebrachium

13. Body area number 7 is called _____.

    answer: popliteal

14. Body area number 8 is called _____.

    answer: sura (sural)

15. Body area number 9 is called _____.

    answer: tarsal

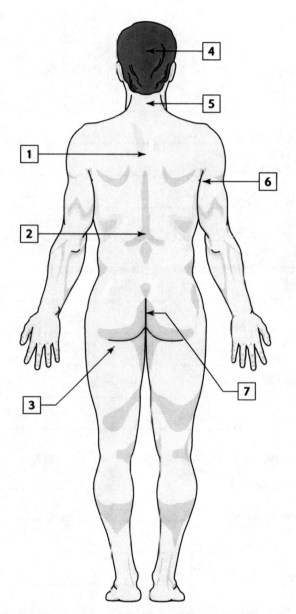

Figure 1-5: Superficial landmarks (posterior view; head and torso superficial terms).

16. Body area number 1 is called _____.

    answer: costal

17. Body area number 2 is called _____.

     answer: lumbar

18. Body area number 3 is called _____.

     answer: gluteal fold

Work Problems

Answer Questions 1 through 5 using a body term from Figure 1-2.

1. Body area number 2 is called _____.

2. Body area number 4 is called _____.

3. Body area number 6 is called _____.

4. Body area number 8 is called _____.

5. Body area number 10 is called _____.

Answer Questions 6 through 10 using a body term from Figure 1-3.

6. Body area number 1 is called _____.

7. Body area number 3 is called _____.

8. Body area number 5 is called _____.

9. Body area number 7 is called _____.

10. Body area number 9 is called _____.

Answer Questions 11 through 15 using a body term from Figure 1-4.

11. Body area number 2 is called _____.

12. Body area number 4 is called _____.

13. Body area number 5 is called _____.

14. Body area number 6 is called _____.

15. Body area number 10 is called _____.

Answer Questions 16 through 19 using a body term from Figure 1-5.

16. Body area number 4 is called _____.

17. Body area number 5 is called _____.

18. Body area number 6 is called _____.

19. Body area number 7 is called _____.

Worked Solutions

1. **antecubital.** The antecubital is the area anterior to the elbow region.

2. **carpal.** The wrist area is called the carpal region.

3. **femoral.** The upper thigh is the femoral region.

4. **crus.** The anterior lower leg is the crus (crural).

5. **hallux.** The big toe is the hallux. The thumb is the pollex.

6. **ocular.** The eye region is called the ocular or oculus.

7. **thoracic.** The chest region is the thoracic.

8. **inguinal.** The groin region is the inguinal.

9. **otic.** The ear region is the otic. The instrument to look inside the ear is called an otoscope.

10. **axilla.** The armpit region is the axilla.

11. **cubital.** The elbow region is the cubital. Some texts refer to this region as the olecranon.

12. **carpal.** This is the wrist area. The posterior view is the same as the anterior view.

13. **pollex.** The thumb is the pollex. The big toe is the hallux.

14. **femoral.** The posterior upper thigh is the femoral just at the anterior upper thigh.

15. **calcaneus.** The heel of the foot is the calcaneus.

16. **occipital.** The back of the head is the occipital.

17. **cervical.** The entire neck region is the cervical.

18. **axilla.** The armpit region is the axilla.

19. **gluteal cleft.** The sagittal crease between the gluteal regions is the gluteal cleft.

Body Quadrants

The torso of the body consists of the thoracic region, the abdominal region, and the pelvic region. The thoracic region consists of the lungs and the heart. The thoracic region and abdominal region are separated by a muscle called the diaphragm muscle. This is our major breathing muscle. The abdominal region and pelvic region are not separated by any physical structure. Therefore, those two regions are combined together to form the abdominopelvic region.

The abdominopelvic region consists of numerous organs. Therefore, anatomists have subdivided the abdominopelvic region into four quadrants to make it easier for the physician to make a preliminary diagnosis.

Look at Figure 1-6 and match the body quadrant name with the appropriate number. The body quadrant names are: right upper quadrant (RUQ), right lower quadrant (RLQ), left upper quadrant (LUQ), and left lower quadrant (LLQ). Keep in mind, when we speak of the right side of the body, it is the patient's right side, not your right side.

After studying Figure 1-6, answer the four questions following the Figure.

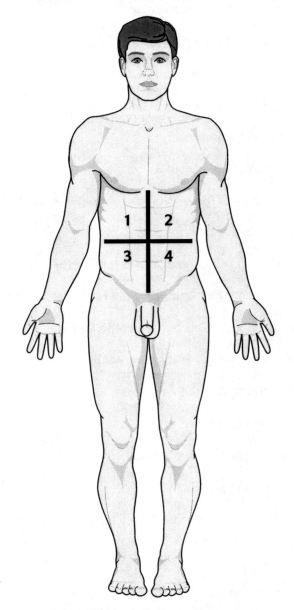

Figure 1-6: Body quadrant.

Example Problems

Answer the following questions using Figure 1-6.

1. Body quadrant 1 is called _____.

   answer: right upper quadrant (RUQ). Remember, the right side of the body is the patient's right side, not yours.

2. Body quadrant 4 is called _____.

   answer: left lower quadrant (LLQ)

3. Most of the liver is located in which body quadrant?

   answer: RUQ. The liver is mostly on the right side of the abdomen region.

4. Most of the stomach is located in which body quadrant?

   answer: LUQ. Most of the stomach is located to the left of the midline of the body.

5. All of the spleen is located in which body quadrant?

   answer: LUQ. The spleen is located on the left side of the stomach.

Abdominopelvic Regions

Because there are so many organs in the abdomen and pelvis (abdominopelvic), many physicians prefer the quadrants to be further subdivided into nine **abdominopelvic regions.** Look at Figure 1-7 to see how the imaginary lines are drawn to create the nine abdominopelvic regions. Table 1-2 lists the terminology associated with the nine abdominopelvic regions.

Table 1-2 Abdominopelvic Regions	
Abdominopelvic Term	*Description of Location*
Umbilical	In the middle of the abdomen
Epigastric	Superior to the umbilical
Hypogastric	Inferior to the umbilical
Left hypochondriac	Left of the epigastric
Right hypochondriac	Right of the epigastric
Left lumbar	Left of the umbilical
Right lumbar	Right of the umbilical
Left inguinal (iliac)	Left of the hypogastric
Right inguinal (iliac)	Right of the hypogastric

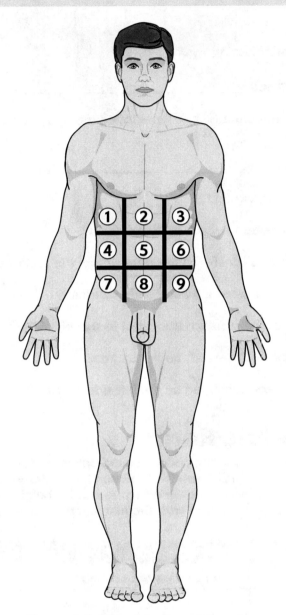

Figure 1-7: Nine abdominopelvic regions.

Example Problems

Use Table 1-2 to match the name of the abdominopelvic region with its location on Figure 1-7.

1. What is the name for abdominopelvic region number 2?

    answer: epigastric

2. What is the name for abdominopelvic region number 3?

    answer: left hypochondriac

3. What is the name for abdominopelvic region number 4?

    answer: right lumbar

4. What is the name for abdominopelvic region number 7?

    answer: right inguinal (iliac)

5. What is the name for abdominopelvic region number 8?

    answer: hypogastric

Physicians are able to use the nine abdominopelvic regions to help them determine what is wrong with a patient when the patient walks into the doctor's office and says, "It hurts right here." The patient then points to an area and based on that location, the doctor can determine which organ of the body may be involved in the pain that the patient feels.

Table 1-3 lists various organs located within the abdominopelvic region. The locations of those organs are briefly described. After examining Table 1-3 and referring to Figure 1-7, answer the example problems that follow.

Table 1-3 Location of Organs in the Abdominopelvic Region	
Abdominopelvic Organ	**Description of Location**
Liver	The biggest part (the rounded part called the fundus) is located on the right side of the body, inferior to the diaphragm muscle. The smallest part of the liver is located left of the fundus.
Stomach	The biggest part (the rounded part called the fundus) is located on the left side of the body, inferior to the diaphragm muscle.
Spleen	Located lateral to the fundus of the stomach.
Ascending colon	Located on the middle right side of the body.
Descending colon	Located on the middle left side of the body.
Cecum	Located on the lower right side, attached to the ascending colon. It is the first part of the large intestine (colon).
Appendix	Connected to the cecum. Located right on the border of two lower abdominopelvic regions.
Urinary bladder	Located in the center lower abdominopelvic region.
Small intestine	Located primarily in the center of the abdomen.

Example Problems

Study the names of the nine abdominopelvic regions. Answer the following questions by using one of the abdominopelvic terms.

1.   The right portion of the liver is located primarily in which of the abdominopelvic regions?

     answer: right hypochondriac

2.   The major portion of the stomach is located primarily in which of the abdominopelvic regions?

     answer: left hypochondriac

3.   The spleen is located in which of the abdominopelvic regions?

     answer: left hypochondriac

4.   The urinary bladder is located in which of the abdominopelvic regions?

     answer: hypogastric

5.   The ascending colon is located in which of the abdominopelvic regions?

     answer: right lumbar

Work Problems

Answer numbers 1–5 after studying Figure 1-6.

1.   According to Figure 1-6, body quadrant 2 is called _____.

2.   According to Figure 1-6, body quadrant 3 is called _____.

3.   The urinary bladder is located in which two body quadrants?

4.   The appendix is located in which body quadrant?

5.   The gallbladder is located in which body quadrant?

Answer numbers 6–10 after studying Figure 1-7.

6.   What is the name for abdominopelvic region number 1?

7.   What is the name for abdominopelvic region number 5?

8.   What is the name for abdominopelvic region number 6?

9.   The lumbar abdominopelvic region is (lateral or medial) to the umbilical region?

10. What is the name for abdominopelvic region number 9?

11. The liver is located primarily to the (left or right) side of the midline of the body.

12. The stomach is located primarily to the (left or right) of the midline of the body.

13. What is the name of the abdominopelvic region that is located in the center of the abdomen?

14. What is the name of the abdominopelvic region that is located in the midline area immediately inferior to the diaphragm muscle?

15. Which abdominal quadrant contains the majority of the stomach?

16. The naval is located in which abdominopelvic region?

17. The appendix is located in which abdominal quadrant?

18. What is the name for the abdominopelvic region that is located left lateral to the hypogastric region?

19. What is the name for the abdominopelvic region that is located medial to the left lumbar region?

20. Are the lungs located in an abdominopelvic region?

Worked Solutions

1. **left upper quadrant (LUQ).** Remember, the left side of the body is the patient's left side, not yours.

2. **right lower quadrant (RLQ)**

3. **RLQ and LLQ**

4. **RLQ**

5. **RUQ**

6. **right hypochondriac.** Remember that the right side of the body is the right side of the patient, not your right side.

7. **umbilical**

8. **left lumbar**

9. **lateral.** If you start out in the umbilical area and move left lateral or right lateral, you will find the left and right lumbar region.

10. **left inguinal (iliac)**

11. **The liver is located primarily to the right of the midline.**

12. **Most of the stomach is located to the left of the midline.**

13. **umbilical**

14. **epigastric**

15. **left upper quadrant**

16. **umbilical**

17. **left lower quadrant**

18. **left inguinal (left iliac)**

19. **umbilical**

20. **No, the lungs are in the thoracic cavity.**

Body Planes

During surgeries or autopsies, the surgeon or examiner needs to cut into the body or cut into organs or tissues. In the following examples, you will find that it is no longer sufficient to say, "I am going to cut this piece of tissue in half." There are three different ways to cut a piece of tissue. Keep in mind that the purpose of the language of anatomy is to develop precise and specific explanations or descriptions.

Table 1-4 lists the terms associated with the body planes (or planes of dissection) and a brief description of each.

Table 1-4 Body Planes	
Body Plane Term	**Description**
Sagittal	This plane divides the body right and left. This term encompasses both mid-sagittal and parasagittal.
Midsagittal	This plane divides the body in equal halves, right and left.
Parasagittal	This plane divides the body right and left but not equally.
Transverse	This plane divides the body superior and inferior.
Frontal	This plane divides the body anterior and posterior.

Example Problems

Use one of the body plane terms to answer the following questions.

1. If a surgeon were to make a cut that extends from the medial side of your patella region to the lateral side of the patella region, it would be a _____ cut.

   answer: transverse. In this situation, the cut is separating superior from inferior.

2. If you were to make a cut that separates the face from the occipital region, it would be a _____ cut.

   answer: frontal. In this situation, the cut is separating anterior from posterior.

3. If you were to make a cut that extends from the knuckle region of your hand to the carpal region, it would be a _____ cut.

    answer: sagittal. If it were exactly in the middle, it would be midsagittal; if it were unequal, it would be parasagittal.

4. If you were to make a cut that extends from the center of the frons, goes between the eyes, and ends at the tip of your nose, it would be a _____ cut.

    answer: sagittal (midsagittal). In this situation, the cut is separating the left from the right.

5. If a surgeon were to make a cut that extends from the left lumbar region to the right lumbar region, it would be a _____ cut.

    answer: transverse. In this situation, the cut is separating superior from inferior.

Work Problems

Fill in the following blanks with one of the three dissectional cuts previously discussed (sagittal, transverse, and frontal).

1. If a surgeon were to remove the lower leg from a patient's body, the cut that would be made would be a _____ cut.

2. If a surgeon were to make an incision that extends from the hip to the center of the knee, the incision would be a _____ cut.

3. If you were to cut your finger lengthwise through the fingernail, you would be making a _____ cut.

4. If you were to make a cut from the frons area that extended inferior through the right eye, it would be a _____ cut.

5. The gluteal cleft is a _____ crease in reference to the entire gluteal region.

6. If you were to make a cut that extends from the left hip region and goes across to the right hip region, it would be a _____ cut.

7. If a surgeon were to make a cut parallel to the diaphragm muscle, it would be a _____ cut.

8. If you had to have a finger amputated, the surgeon would typically make a _____ cut.

9. If you were to cut your hand and the cut extended from the medial side of the palm to the thumb, it would be a _____ cut.

10. The heart is located in which of the abdominopelvic regions?

Worked Solutions

1. **This type of cut would be transverse.** By removing the lower leg, the surgeon is separating the inferior from the superior.

2. **This type of cut would be a sagittal cut.** If it were exactly in the middle, it would be midsagittal. If it were not in the exact middle, it would be parasagittal.

3. **This type of cut would be sagittal.** If it were exactly in the middle, it would be midsagittal; if it were not in the exact middle, it would be parasagittal.

4. **This type of cut would be sagittal.** Because this cut goes through the right eye, it is not in the center of the body. While it is separating the right from the left, it would be considered sagittal. However, because it is not in the center of the body, it would be more accurately described as parasagittal.

5. **The gluteal cleft is a sagittal structure.** Because it is the middle of the gluteal region, it can be considered midsagittal.

6. **This type of cut would be transverse.** By making an incision in this manner, the inferior is separated from the superior.

7. **This type of cut would be transverse.** The diaphragm muscle is located inferior to the lungs and superior to the liver and extends from one side of the body to the other side.

8. **This type of cut would be transverse.**

9. **This type of cut would be transverse.** In this situation, the cut is separating superior from inferior.

10. **The heart isn't located in an abdominopelvic region.** The heart is located in the thoracic area (superior to the diaphragm muscle).

Homeostasis

A major physiological theme for physiology courses is understanding the concept of homeostasis. **Homeostasis** is the condition where the body is healthy. There are many mechanisms in the body that are at work all the time to help the body maintain homeostasis. If those mechanisms fail, the body will be out of homeostasis and the individual will need to seek medical attention from a doctor. The doctor then works to get the body back into homeostasis.

There are two major mechanisms. One mechanism is at work 24 hours a day, 7 days a week. The other mechanism is at work only under specific circumstances. Those mechanisms are called the **negative feedback mechanism** and **positive feedback mechanism,** respectively.

Negative Feedback Mechanism

The negative feedback mechanism works every minute of the day to help us maintain homeostasis. The negative feedback mechanism works on a fluctuation principle. For example, when our body temperature begins to rise, there are mechanisms in place that begin to bring the temperature back down to a normal level. If our body temperature begins to fall, the mechanisms begin to bring the temperature back up to a normal level.

It is very important that the level of calcium ions in our blood is normal. Whenever the calcium ions begin to fall, there are mechanisms in place that begin to bring the calcium ions back up to their normal level.

Positive Feedback Mechanism

The positive feedback mechanism works only under certain conditions. The positive feedback mechanism does not exhibit any kind of fluctuation. For example; if a person has a bacterial infection, their body temperature begins to climb. Many species of bacteria will die when exposed to high temperatures. If the body senses that not all the bacteria are dead, it will cause the temperature to rise some more. If it still senses that all the bacteria are not dead, it will cause the temperature to rise once again. Pretty soon, the temperature is up to 103°F. This of course is a fever. Notice in this example, the temperature did not fluctuate. It continued in one direction until it accomplished its goal: to kill bacteria.

Example Problems

Identify the type of mechanism being described.

1. When the blood begins to clot, the chemical reactions involved in the process will continue until the clot is completed.

    answer: positive feedback. The blood must continue the clotting process until the clot is completely formed.

2. The level of insulin will rise under certain conditions but will soon drop to normal levels in the presence of another hormone called glucagon. This process typically occurs throughout the day.

    answer: negative feedback. Because the level of insulin rises and then falls (fluctuates) all day long, it is called a negative feedback mechanism.

3. Blood vessels will automatically dilate in an effort to release body heat and they will constrict in an effort to retain body heat. This is done to try to maintain the correct body temperature.

    answer: negative feedback. Because the blood vessels dilate and constrict, they are changing size in a fluctuating manner. A fluctuating system is due to negative feedback mechanisms.

4. Once in awhile, there is a chemical reaction that produces a product that will enhance the chemical reaction even more and therefore producing even more of the product.

    answer: positive feedback. Because one product causes the original substance to react even more, the system is not fluctuating. Because the system is not fluctuating, a positive feedback mechanism is at work.

5. The temperature of your house cools down, which triggers the heat mechanism to warm up the house. As soon as the house gets too warm, the heater shuts off to ultimately cool the house down. This process repeats.

    answer: negative feedback. The temperature of the house fluctuates. It is regulated by a negative feedback mechanism.

Chapter Problems and Solutions

Problems

For problems 1–10, use superficial body terms.

1. Identify the superficial term that refers to the upper arm.

2. When most people speak about the elbow, they are typically referring to the posterior side. Identify the superficial term that refers to the elbow.

3. Identify the superficial term that refers to the area that is posterior to the patella.

4. The wrist area is called the _____ and the ankle area is called the _____.

5. The ears are called the _____.

6. The anterior lower leg is called the _____ and the posterior lower leg is called the _____.

7. The superficial area that is posterior to the frons is the _____.

8. The area immediately inferior to the oris is the _____.

9. The area superior to the diaphragm muscle (also superior to the abdominal area) is the _____.

10. The upper back is called the _____ and the lower back is called the _____.

For problems 11–20, use directional terminology.

11. The thumbs are _____ structures.

12. The palms of the hands face _____ according to the anatomical position.

13. The left side of the right leg is _____ to the right side of the right leg.

14. The popliteal is a (an) _____ structure.

15. The antecubital is a (an) _____ structure.

16. The ears are a (an) _____ structure.

17. The naval is located on the _____ side of the body.

18. The bulgy part of the elbow (cubital) is located on the _____ of the arm.

19. The first finger (digit number 2) is located _____ to the thumb.

20. The hallux is located _____ to the little toe.

For questions 21–25, use quadrant terms and/or abdominopelvic terms.

21. The majority of the liver is located in the _____ quadrant but more specifically in the _____ region.

22. The fundus of the stomach is located in the _____ quadrant but more specifically in the _____ region.

23. The urinary bladder is located on the border of the _____ and _____ quadrant but more specifically in the _____ region.

24. The naval is located right at the junction of all four quadrants but more specifically in the _____ region.

25. Parts of the small intestine can be found in all of the quadrants but most of it is found specifically in the _____ region.

For questions 26–30, use body plane terminology.

26. If you were to make a cut that extended from the cubital region to the posterior carpal region, you would be making a _____ cut.

27. If you were to make a cut that extended from the popliteal region to the patellar region, you would be making a _____ cut.

28. If you were to make a cut that extended the length of the finger but yet left the fingernail intact, you would be making a _____ cut.

29. If you were to make a cut that extended from the naval to the mentis, you would be making a _____ cut.

30. If you were to make a cut that extended from the popliteal to the calcaneus, you would be making a _____ cut.

Answers and Solutions

1. **brachium.** This term refers to the upper arm.

2. **cubital (olecranon).** This term refers to the elbow region.

3. **popliteal.** This term refers to the posterior side of the knee.

4. **carpal; tarsal**

5. **otic.** This term refers to the ear region.

6. **crus (crural); sura (sural).** The anterior lower leg is the crus while the posterior lower leg is the sura.

7. **occipital.** This term refers to the back of the head.

8. **mentis.** This term refers to the chin area.

9. **thoracic.** This term refers to the chest area.

10. **costal; lumbar.** The upper back is the costal region (sometimes called the dorsal region) and the lower back is the lumbar region.

11. **lateral.** In the anatomical position, the thumbs are farthest from the midline of the body.

12. **anterior.** In the anatomical position, the palms are facing forward.

13. **medial.** The left side of the right leg is medial. In layman's terms it is the inside of the right leg. Anatomists only use the term inside when they are literally looking "inside" the body.

14. **posterior.** The popliteal is on the "back side" of the knee.

15. **anterior.** The antecubital is on the "front side" of the arm.

16. **lateral.** The ears are on the "side" of the head, away from the midline.

17. **anterior.** The naval is the "belly button" and is on the "front side" of the body.

18. **posterior.** The cubital is the elbow region and is on the "back side" of the arm.

19. **medial.** In the anatomical position, if you move from the thumb to the first finger (digit 2) you have to move medial (closer to the midline of the body).

20. **medial.** The hallux is the big toe and the big toe is closer to the midline of the body than the little toe.

21. **RUQ; right hypochondriac.** Most of the liver is on the right side of the abdomen.

22. **LUQ; left hypochondriac.** The fundus is the rounded part of the stomach and is located mainly on the left side of the abdomen.

23. **LLQ and RLQ; hypogastric.** The urinary bladder is located between the inguinal regions.

24. **umbilical.** The naval is the "belly button" and is located in the center of the abdomen.

25. **umbilical.** If you put your hand on your naval, your hand would be resting on most of the small intestine.

26. **sagittal.** If it were exactly in the middle, it would be midsagittal; if it were unequal, it would be parasagittal.

27. **transverse.** You are separating the superior from the inferior.

28. **frontal.** You are separating the anterior from the posterior.

29. **midsagittal.** You are separating the right from the left.

30. **sagittal.** If it were exactly in the middle, it would be midsagittal; if it were unequal, it would be parasagittal.

Supplemental Chapter Problems

Problems

1.   Which directional term is used when we are moving from the cubital region to the shoulder area?

2.   Which directional term is used when we are moving from the left side of the right arm to the right side of the right arm?

3.   Which two directional terms are used when we are moving from the umbilical region to the left inguinal region?

4.   Which directional term is used when we are moving from the top of the ear to the ear lobe?

5.   Which directional term is used when we are moving from the RUQ to the LLQ?

6.   Identify an organ that is located in the right hypochondriac region.

7.   Identify an organ that is located in the left hypochondriac region.

8.   Identify an organ that is located in the hypogastric region.

9.   Which quadrant is the appendix located in?

10.   The thymus gland is located slightly superior to the heart. It is located in which abdominopelvic region?

11.   If a surgeon were to make an incision that parallels the diaphragm muscle, he/she would be making a _____ cut.

12.   If you were to make a cut that extended from the costal region to the lumbar region that equally divided the portions, you would be making a _____ cut.

13.   The following is a picture of a heart. If you were to make an incision along line A, you would be making a _____ cut.

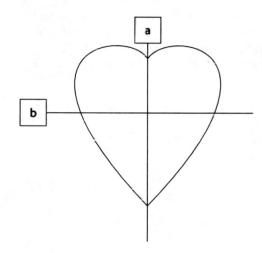

14. If you were to make an incision along line B in the heart picture in number 13, you would be making a _____ cut.

15. Examine the heart picture in number 13. Explain why we have to use the body plane terminology when describing various cuts on organs or tissues such as the heart. Why can't we simply say, "We are going to cut the heart in half?"

Answers

1. superior

2. lateral

3. inferior and left lateral or left lateral and then inferior

4. inferior

5. inferior and left lateral or left lateral and then inferior

6. liver

7. stomach and/or spleen

8. urinary bladder

9. RLQ

10. The thymus gland is in the thoracic region, not in the abdominopelvic regions.

11. transverse

12. midsagittal

13. midsagittal

14. transverse

15. Based on the diagram of the heart, there are two ways to cut the heart in half. You can have it equally divided from left and right (midsagittal) and you can also have it equally divided superior from inferior (transverse). You can also have it equally divided anterior from posterior (frontal), but this view isn't shown.

Chapter 2

The Chemistry of Anatomy and Physiology

Through the study of anatomy and physiology, the subject of chemistry comes up quite often. Chemical reactions are necessary for the building of bone material, muscle contraction, developing immunity to chicken pox, simple breathing, and so on. This chapter helps you to review some chemistry concepts, ranging from the basic atom to complex proteins.

Atoms

The atom consists of a **nucleus** (not to be confused with the nucleus of a cell, which is discussed in Chapter 3) and orbits. The atom is made of three subatomic particles. They are the **proton, electron,** and **neutron.**

Table 2-1 lists some facts concerning the three subatomic particles.

Table 2-1 Atom Facts	
Subatomic Particle	*General Facts*
Proton	The number of protons an atom has gives it its identity.
	The proton number never changes.
	The number of protons plus the number of neutrons gives the atom its mass value (atomic mass unit or AMU).
	Protons have a positive charge.
Neutron	The number of neutrons can change.
	When this number changes, the atom becomes an isotope.
	Many times, these isotopes are radioactive.
	Use this equation to determine the number of neutrons an atom has: AMU – proton number = neutron number
	Neutrons do not have a charge.
Electrons	Electrons orbit around the nucleus.
	A neutral atom has the same number of electrons as it has protons.
	Electrons have a negative charge.

The Periodic Table

To make your understanding of chemistry easier, it is best to develop an understanding of how to use the Periodic Table (see Figure 2-1). Knowing how to read the Periodic Table will provide numerous avenues to the understanding of chemistry. The Periodic Table was designed in such a manner so that the atoms in the first column have some common characteristics and the atoms in the second column also have some things in common but yet are different than the atoms in column one.

For example, all the atoms in the same column as hydrogen (H) will develop a positive one charge when they become ions. Once they become ions they will be written in this manner; H^{1+}, Na^{1+}, and so on. All the atoms in the same column as beryllium (Be) will develop a positive two charge when they become ions. Once they become ions they will be written in this manner; Be^{2+}, Ca^{2+}, and so on. All the atoms in the same column as boron (B) will develop a positive three charge. The atoms in the column with nitrogen (N) will develop a negative three charge, the column with oxygen (O) will develop a negative two charge, and the column with fluorine (F) will develop a negative one charge.

The transition elements (Sc through Zn) don't seem to have a nice pattern, and those in the very last column with helium (He) do not become ions. Keep in mind that there are exceptions.

PERIODIC TABLE OF THE ELEMENTS

1 H Hydrogen 1.00797																	2 He Helium 4.0026
3 Li Lithium 6.939	4 Be Beryllium 9.0122											5 B Boron 10.811	6 C Carbon 12.01115	7 N Nitrogen 14.0067	8 O Oxygen 15.9994	9 F Flourine 18.9984	10 Ne Neon 20.183
11 Na Sodium 22.9898	12 Mg Magnesium 24.312											13 Al Aluminum 26.9815	14 Si Silicon 28.086	15 P Phosphorus 30.9738	16 S Sulfur 32.064	17 Cl Chlorine 35.453	18 Ar Argon 39.948
19 K Potassium 39.102	20 Ca Calcium 40.08	21 Sc Scandium 44.956	22 Ti Titanium 47.90	23 V Vanadium 50.942	24 Cr Chromium 51.996	25 Mn Manganese 54.9380	26 Fe Iron 55.847	27 Co Cobalt 58.9332	28 Ni Nickel 58.71	29 Cu Copper 63.546	30 Zn Zinc 65.37	31 Ga Gallium 69.72	32 Ge Germanium 72.59	33 As Arsenic 74.9216	34 Se Selenium 78.96	35 Br Bromine 79.904	36 Kr Krypton 83.80
37 Rb Rubidium 85.47	38 Sr Strontium 87.62	39 Y Yttrium 88.905	40 Zr Zirconium 91.22	41 Nb Niobium 92.906	42 Mo Molybdenum 95.94	43 Tc Technetium (99)	44 Ru Ruthenium 101.07	45 Rh Rhodium 102.905	46 Pd Palladium 106.4	47 Ag Silver 107.868	48 Cd Cadmium 112.40	49 In Indium 114.82	50 Sn Tin 118.69	51 Sb Antimony 121.75	52 Te Tellurium 127.60	53 I Iodine 126.9044	54 Xe Xenon 131.30
55 Cs Cesium 132.905	56 Ba Barium 137.34	57 La Lanthanum 138.91	72 Hf Hafnium 179.49	73 Ta Tantalum 180.948	74 W Tungsten 183.85	75 Re Rhenium 186.2	76 Os Osmium 190.2	77 Ir Iridium 192.2	78 Pt Platinum 195.09	79 Au Gold 196.967	80 Hg Mercury 200.59	81 Tl Thallium 204.37	82 Pb Lead 207.19	83 Bi Bismuth 208.980	84 Po Polonium (210)	85 At Astatine (210)	86 Rn Radon (222)
87 Fr Francium (223)	88 Ra Radium (226)	89 Ac Actinium (227)	104 Rf Rutherfordium (261)	105 Db Dubnium (262)	106 Sg Seaborgium (266)	107 Bh Bohrium (264)	108 Hs Hassium (269)	109 Mt Meitnerium (268)	110 Uun Ununnilium (269)	111 Uuu Unununium (272)	112 Uub Ununbium (277)	113 Uut §	114 Uuq Ununquadium (285)	115 Uup §	116 Uuh Ununhexium (289)	117 Uus §	118 Uuo Ununoctium (293)

Lanthanide Series	58 Ce Cerium 140.12	59 Pr Praseodymium 140.907	60 Nd Neodymium 144.24	61 Pm Promethium (145)	62 Sm Samarium 150.35	63 Eu Europium 151.96	64 Gd Gadolinium 157.25	65 Tb Terbium 158.924	66 Dy Dysprosium 162.50	67 Ho Holmium 164.930	68 Er Erbium 167.26	69 Tm Thulium 168.934	70 Yb Ytterbium 173.04	71 Lu Lutetium 174.97
Actinide Series	90 Th Thorium 232.038	91 Pa Protactinium (231)	92 U Uranium 238.03	93 Np Neptunium (237)	94 Pu Plutonium (242)	95 Am Americium (243)	96 Cm Curium (247)	97 Bk Berkelium (247)	98 Cf Californium (251)	99 Es Einsteinium (254)	100 Fm Fermium (257)	101 Md Mendelevium (258)	102 No Nobelium (259)	103 Lr Lawrencium (260)

§ Note: Elements 113, 115, and 117 are not known at this time, but are included in the table to show their expected positions.

Figure 2-1: The Periodic Table.

Example Problems

1. Which subatomic particle is found in the orbital rings of an atom?

   answer: electrons

2. The electrons associated with an atom are orbiting around the _____.

   answer: nucleus

3. When calcium becomes an ion, how would it be written?

   answer: Ca^{2+}. Because calcium is found in the second column on the Periodic Table, it therefore develops a 2+ charge.

4. How many neutrons does strontium have?

   answer: 50 neutrons. The AMU value for Sr is 88. It has 38 protons. $88 - 38 = 50$.

5. How many protons does oxygen have?

   answer: 8. The number 8 on the Periodic Table associated with oxygen is the number of protons oxygen has.

Ions

Ions are atoms that have either gained or lost electrons. When they gain or lose electrons, they develop a charge. When they gain electrons, they develop a negative charge. When they lose electrons, they develop a positive charge. When atoms become ions, they can become quite reactive. When they are reactive, they bond together, thus creating new substances. To understand how the ions bond together, refer to the Periodic Table. Examine the following examples.

Example 1: Bond calcium ions to phosphorus ions.

1. Calcium ions are written as Ca^{2+}.
2. Phosphorus ions are written as P^{3-}.
3. Write the ions in this manner: $Ca^{2+}P^{3-}$.
4. Take the 3 (from 3−) and write it next to Ca in the lower right region (Ca_3).
5. Take the 2 (from 2+) and write it next to P in the lower right region (P_2).
6. The product will now look like this: Ca_3P_2.

Example 2: Bond sodium ions to oxygen ions.

1. Sodium ions are written as Na^{1+}.
2. Oxygen ions are written as O^{2-}.
3. Write the ions in this manner: $Na^{1+}O^{2-}$.
4. Take the 2 (from 2−) and write it next to Na in the lower right region (Na_2).
5. Take the 1 (from 1+) and write it next to O in the lower right region (O_1).
6. The product will now look like this: Na_2O_1 or it can also be written as Na_2O.

Example Problems

1. Bond lithium ions to sulfur ions.

    answer: Li_2S_1 or Li_2S

    When lithium becomes an ion, it will develop a charge of 1+ (it is located in the first column on the periodic chart). When sulfur becomes an ion, it will develop a charge of 2– (it is located in the column under oxygen).

    Begin by writing the product as: $Li^{1+}S^{2-}$.

    Place the 2 (from sulfur) in the subscript form with Li (Li_2).

    Place the 1 (from lithium) in the subscript form with S (S_1).

2. Bond magnesium ions to bromine ions.

    answer: Mg_1Br_2 or $MgBr_2$

    When magnesium becomes an ion, it will develop a charge of 2+ (it is located in the second column on the periodic chart). When bromine becomes an ion, it will develop a charge of 1– (it is located in the column under fluorine).

    Begin by writing the product as: $Mg^{2+}Br^{1-}$.

    Place the 1 (from bromine) in the subscript form with Mg (Mg_1).

    Place the 2 (from magnesium) in the subscript form with Br (Br_2).

3. Bond potassium ions to chlorine ions.

    answer: K_1Cl_1 or KCl

    When potassium becomes an ion, it will develop a charge of 1+ (it is located in the first column on the periodic chart). When chlorine becomes an ion, it will develop a charge of 1– (it is located in the column under fluorine).

    Begin by writing the product as: $K^{1+}Cl^{1-}$.

    Place the 1 (from chlorine) in the subscript form with K (K_1).

    Place the 1 (from potassium) in the subscript form with Cl (Cl_1).

4. Bond aluminum ions to chlorine ions.

    answer: Al_1Cl_3 or $AlCl_3$

    When aluminum becomes an ion, it will develop a charge of 3+ (it is located in the third major column on the periodic chart). When chlorine becomes an ion, it will develop a charge of 1– (it is located in the column under fluorine).

Begin by writing the product as: $Al^{3+}Cl^{1-}$.

Place the 1 (from chlorine) in the subscript form with Al (Al_1).

Place the 3 (from aluminum) in the subscript form with Cl (Cl_3).

5. Bond aluminum ions to sulfur ions.

    answer: Al_2S_3

    When aluminum becomes an ion, it will develop a charge of 3+ (it is located in the third major column on the periodic chart). When sulfur becomes an ion, it will develop a charge of 2– (it is located in the column under oxygen).

    Begin by writing the product as: $Al^{3+}S^{2-}$.

    Place the 2 (from sulfur) in the subscript form with Al (Al_2).

    Place the 3 (from aluminum) in the subscript form with S (S_3).

Polyatomic Ions

The ions previously studied are single ions. They are single atoms with a single charge. There is another group of ions called **polyatomic ions.** Polyatomic ions are several atoms having a single charge. Examples of polyatomic ions are:

- ❏ NH_4^{1+} Ammonium ion
- ❏ OH^{1-} Hydroxide ion
- ❏ HCO_3^{1-} Bicarbonate ion
- ❏ CO_3^{2-} Carbonate ion
- ❏ PO_4^{3-} Phosphate ion

Polyatomic ions can be bonded to other polyatomic ions or single ions just as two single ions can bond together.

Example 1: Bond ammonium ions to phosphorus ions.

1. Ammonium ions are written as NH_4^{1+}.
2. Phosphorus ions are written as P^{3-}.
3. Write the ions in this manner: $NH_4^{1+}P^{3-}$.
4. Take the 3 (from 3–) and write it next to NH_4 in the lower right region [$(NH_4)_3$].
5. Take the 1 (from 1+) and write it next to P in the lower right region (P_1).
6. The product will now look like this: $(NH_4)_3P_1$ or $(NH_4)_3P$.

    It is important to remember to put parenthesis around the polyatomic ion because the equation indicates that you need 3 ammonium ions and each ammonium ion is written as NH_4.

Example 2: Bond aluminum ions to carbonate ions.

1. Aluminum ions are written as Al^{3+}.
2. Carbonate ions are written as CO_3^{2-}.

3. Write the ions in this manner: $Al^{3+}CO_3{}^{2-}$.

4. Take the 2 (from 2−) and write it next to Al in the lower right region Al_2.

5. Take the 3 (from 3+) and write it next to CO_3 in the lower right region $[(CO_3)_3]$.

6. The product will now look like this: $Al_2(CO_3)_3$.

It is important to remember to put parenthesis around the polyatomic ion because the equation indicates that you need 3 carbonate ions and each carbonate ion is written as CO_3.

Example Problems

1. Bond lithium ions to carbonate ions.

   answer: $Li_2(CO_3)_1$ or Li_2CO_3

   When lithium becomes an ion, it will develop a charge of 1+ (it is located in the first column on the periodic chart). The carbonate ion has a charge of 2−.

   Begin by writing the product as: $Li^{1+}CO_3{}^{2-}$.

   Place the 2 (from the carbonate ion) in the subscript form with Li (Li_2).

   Place the 1 (from the lithium ion) in the subscript form with CO_3 $[(CO_3)_1]$.

2. Bond magnesium ions to bicarbonate ions.

   answer: $Mg_1(HCO_3)_2$ or $Mg(HCO_3)_2$

   When magnesium becomes an ion, it will develop a charge of 2+ (it is located in the second major column on the periodic chart). The bicarbonate ion has a charge of 1−.

   Begin by writing the product as: $Mg^{2+}HCO_3{}^{1-}$.

   Place the 1 (from the bicarbonate ion) in the subscript form with Mg (Mg_1).

   Place the 2 (from magnesium) in the subscript form with HCO_3 $[(HCO_3)_2]$.

3. Bond potassium ions to phosphate ions.

   answer: $K_3(PO_4)_1$ or K_3PO_4

   When potassium becomes an ion, it will develop a charge of 1+ (it is located in the first column on the periodic chart). The phosphate ion has a charge of 3−.

   Begin by writing the product as: $K^{1+}PO_4{}^{3-}$.

   Place the 3 (from the phosphate ion) in the subscript form with K (K_3).

   Place the 1 (from potassium) in the subscript form with PO_4 $[(PO_4)_1]$.

4. Bond ammonium ions to oxygen ions.

   answer: $(NH_4)_2O$

The ammonium ion has a charge of 1+. When oxygen becomes an ion, it will develop a charge of 2– (elements in the column with oxygen develop a 2– charge).

Begin by writing the product as: $NH_4^{1+}O^{2-}$.

Place the 2 (from oxygen) in the subscript form with NH_4 [$(NH_4)_2$].

Place the 1 (from the ammonium ion) in the subscript form with O (O_1).

5. Bond ammonium ions to carbonate ions.

   answer: $(NH_4)_2CO_3$

   The ammonium ion has a charge of 1+. The carbonate ion has a charge of 2–.

   Begin by writing the product as: $NH_4^{1+}CO_3^{2-}$.

   Place the 2 (from the carbonate ion) in the subscript form with NH_4 [$(NH_4)_2$].

   Place the 1 (from the ammonium ion) in the subscript form with CO_3 [$(CO_3)_2$].

Isotopes

Atoms having the same number of protons but a different number of neutrons are considered to be **Isotopes** of each other. To determine the number of neutrons an atom has, perform the following mathematics:

Atomic mass unit of the atom minus the number of protons equals the number of neutrons.

The significance of having this knowledge is the fact that many times when an atom has an altered number of neutrons, it becomes radioactive. When this happens, it is called a radioactive isotope. Radioactive isotopes can be used in the field of medicine.

The following are examples regarding how to calculate neutron numbers.

Example 1: Calculate the number of neutrons sodium (Na) has.

1. Look at the Periodic Table to determine the atomic mass unit for sodium (23).
2. Look at the Periodic Table to determine the number of protons sodium has (11).
3. Determine the number of neutrons by using this equation:
   AMU – proton number = neutron number (23 – 11 = 12).
4. Sodium has 12 neutrons.

Example 2: Calculate the number of neutrons phosphorus (P) has.

1. Look at the Periodic Table to determine the atomic mass unit for phosphorus (31).
2. Look at the Periodic Table to determine the number of protons phosphorus has (15).
3. Determine the number of neutrons by using this equation:
   AMU – proton number = neutron number (31 – 15 = 16).
4. Phosphorus has 16 neutrons.

Example Problems

Calculate the number of neutrons the following atoms have.

1. Lithium atom (Li)

    answer: 4 neutrons. According to the Periodic Table, lithium has an AMU of 7. It has 3 protons. 7 − 3 = 4

2. Calcium atom (Ca)

    answer: 20 neutrons. According to the Periodic Table, calcium has an AMU of 40. It has 20 protons. 40 − 20 = 20

3. Hydrogen atom (H)

    answer: 0 neutrons. According to the Periodic Table, hydrogen has an AMU of 1. It has 1 proton. 1 − 1 = 0

4. Phosphorus atom (P)

    answer: 16 neutrons. According to the Periodic Table, phosphorus has an AMU of 31. It has 15 protons. 31 − 15 = 16.

5. Potassium atom (K)

    answer: 20 neutrons. According to the Periodic Table, potassium has an AMU of 39. It has 19 protons. 39 − 19 = 20.

Radioactive Isotopes

When you know the number of neutrons an atom has, you can change that number and make the atom become a **radioactive isotope.** Notice that whenever the neutron number changes, the atomic mass unit automatically changes as well (remember, the proton number never changes).

In order to designate which isotope is being discussed, chemists write the isotopes in this manner: ^{23}Na and ^{24}Na. The superscript numbers represent the atomic mass unit value for the atom. ^{23}Na (read as sodium 23) has an atomic mass unit of 23 and has 12 neutrons and 11 protons. Sodium 24 has an atomic mass unit of 24 and has 13 neutrons and 11 protons.

Example Problems

For the following problems, calculate the number of neutrons each isotope has and indicate their written designation.

1. Potassium 39

    answer: 20 neutrons, ^{39}K. According to this problem, this isotope of potassium has an AMU of 39. Potassium has 19 protons. 39 − 19 = 20.

2. Potassium 41

    answer: 22 neutrons, ^{41}K. According to this problem, this isotope of potassium has an AMU of 41. Potassium has 19 protons. 41 − 19 = 22.

3. Technetium 98

    answer: 55 neutrons, ^{98}Tc. According to this problem, this isotope of technetium has an AMU of 98. Technetium has 43 protons. 98 – 43 = 55.

4. Technetium 99

    answer: 56 neutrons, ^{99}Tc. According to this problem, this isotope of technetium has an AMU of 99. Technetium has 43 protons. 99 – 43 = 56.

5. Carbon 12

    answer: 6 neutrons, ^{12}C. According to this problem, this isotope of carbon has an AMU of 12. Carbon has 6 protons. 12 – 6 = 6.

6. Carbon 14

    answer: 8 neutrons, ^{14}C. According to this problem, this isotope of carbon has an AMU of 14. Carbon has 6 protons. 14 – 6 = 8.

Work Problems

Use the Periodic Table to answer Questions 1 through 4.

1. How many protons does sodium have?

2. How many electrons does neutral sodium have?

3. How many neutrons does magnesium 24 have?

4. How many neutrons does magnesium 21 have?

Bond the following ions and polyatomic ions together in Questions 5 through 8.

5. Lithium ion + Oxygen ion:

6. Calcium ion + Phosphorus ion:

7. Ammonium ion + Sulfur ion:

8. Calcium ion + Phosphate ion:

For Questions 9 and 10, indicate how to write the isotope.

9. Sodium with an AMU of 25.

10. Potassium with an AMU of 42.

Worked Solutions

1. **According to the Periodic Table, Na (sodium) has 11 protons.**

2. **Sodium has the same number of electrons as protons when it is neutral (no overall charge).** It, therefore, has 11 electrons.

3. **Magnesium 24 has 12 neutrons.** Magnesium (Mg) has an AMU of 24 and has 12 protons. $24 - 12 = 12$. It is only coincidental that it has the same number of neutrons as protons.

4. **Magnesium 21 is an isotope that has an AMU of 21.** Because it is magnesium, it still has 12 protons. Therefore, $21 - 12 = 9$. Magnesium 21 has only 9 neutrons.

5. **According to the Periodic Table, Li would develop a charge of 1+.** It is in the first column. Oxygen is in the column that develops a 2– charge. The result would be: Li_2O_1 or Li_2O.

6. **According to the Periodic Table, Ca would develop a charge of 2+.** It is in the second column. Phosphorus is in the column under nitrogen and therefore would develop a 3– charge. The result would be: Ca_3P_2.

7. **The ammonium ion has a charge of 1+.** According to the Periodic Table, when sulfur becomes an ion, it will develop a charge of 2–. The result would be $(NH_4)_2S$. Parentheses have to be around the ammonium ion because we need to have two of them.

8. **According to the Periodic Table, Ca would develop a charge of 2+.** The phosphate ion has a charge of 3–. The result would be $Ca_3(PO_4)_2$. Parentheses have to be around the phosphate ion because we need to have two of them.

9. **^{25}Na.** The AMU value is written in the superscript left side of the symbol. Notice, the AMU of sodium, on the Periodic Table, is 23. Sodium 25 is an isotope.

10. **^{42}K.** The AMU value is written in the superscript left side of the symbol. Notice, the AMU of potassium, on the Periodic Table, is 39. Potassium 42 is an isotope.

Organic Molecules

There are two major classes of molecules in the human body. One is called **inorganic** molecules. Inorganic molecules were discussed in the preceding section. This section discusses **organic** molecules. Organic molecules are molecules that typically consist of carbon atoms. These molecules are larger than inorganic molecules. There are four major classes of organic molecules: carbohydrates, lipids, proteins, and nucleic acids.

The atoms associated with organic molecules are bonded together by covalent bonds. **Covalent bonds** are typically represented by dashed lines. Each dashed line represents a single bond. Most organic molecules consist of carbon, hydrogen, oxygen, and nitrogen. Table 2-2 lists some common facts about covalent bonds.

Table 2-2 Covalent Bonds	
Atom	*Number of Bonds*
Hydrogen	1
Oxygen	2
Nitrogen	3
Carbon	4

After bonding atoms together via covalent bonds, thus forming an organic molecule, count the number of bonds associated with each atom. If the number of bonds correlates with Table 2-2, the written molecule is correct. In order for the molecule to be stable, it must have the correct number of bonds. Look at the following examples;

$$H \!-\! O \!-\! H$$

1. Each hydrogen has one bond associated with it.

2. The oxygen has two bonds associated with it.

$$O \!=\! C \!=\! O$$

3. Each oxygen has two bonds associated with it.

4. The carbon has four bonds associated with it.

Example Problems

Put the correct number of bonds between the atoms in the following organic molecules.

1. N N

   answer: N≡N Each nitrogen has to have 3 bonds attached to it. This is called a triple bond.

2. O O

   answer: O=O Each oxygen has to have 2 bonds attached to it. This is called a double bond.

3. H C O H

       O

   answer:
   $$H \!-\! C \!-\! O \!-\! H$$
   $$\|$$
   $$O$$

   Each hydrogen has to have one bond attached to it. Each oxygen has to have two bonds attached to it. One oxygen has two single bonds and the other oxygen has a double bond. The carbon has to have a total of four bonds attached to it. This carbon has two single bonds and one double bond.

4.   H

   H C H

       H

   answer:
   $$H$$
   $$|$$
   $$H \!-\! C \!-\! H$$
   $$|$$
   $$H$$

Each hydrogen has to have one bond attached to it. The carbon has to have four bonds attached to it.

5.   H   N   C   O

    answer: H —— N ══ C ══ O

The hydrogen has to have one bond attached to it. The nitrogen has to have three bonds attached to it. In this case, it has one single bond and one double bond. The carbon has to have a total of four bonds. In this case, it has two double bonds. Oxygen has to have two bonds. In this case it has one double bond.

Carbohydrates

Carbohydrates are made of sugar units called **saccharides.** If the carbohydrate is made of one glucose unit, it is called a **monosaccharide.** If it is made of two units, it is a **disaccharide.** If it is made of several units, it is a polysaccharide. Table 2-3 lists facts regarding typical carbohydrate molecules.

Table 2-3 Carbohydrate Facts		
Monosaccharides	*Disaccharides*	*Polysaccharides*
Glucose	Maltose	Starch
Fructose	Sucrose	Glycogen
Galactose	Lactose	

Glucose bonded to glucose (two sugar units) will produce maltose. Glucose bonded to fructose will produce sucrose. Glucose bonded to galactose will produce lactose. Several glucose units bonded together will produce starch or glycogen.

Lipids

In order for an organic molecule to be classified as a lipid, it must be insoluble in water. Most lipids are made of a glycerol molecule and fatty acids. There are four major types of lipids. Table 2-4 lists some facts about lipids.

Table 2-4 Lipid Facts			
Phospholipids	*Glycolipids*	*Fats*	*Cholesterol*
Made of:	Made of:	Made of:	Made of:
One glycerol	One glycerol	One glycerol	Four carbon rings
Two fatty acids	Two fatty acids	Three fatty acids	No glycerol
One phosphate ion	One carbohydrate		No fatty acids
Found in cell membranes	Found in cell membranes	Used to insulate organs of the body	Found in cell membranes
			Needed to make hormones such as testosterone and estrogen

Proteins

Proteins are organic molecules that are made of units of **amino acids.** The individual amino acids are made of carbon, hydrogen, oxygen, and nitrogen atoms. There are over 300 different amino acids in the living world. Of those 300 + amino acids, the human body uses only 20 of them. Those 20 different amino acids are bonded together (via the action of ribosomes) in different combinations to create over 100,000 different protein molecules.

While all the proteins are special in their own way, there is one very special type of protein called an enzyme. An enzyme is a protein molecule that catalyses the chemical reactions in the body. When an enzyme is involved in a chemical reaction, it speeds up the chemical reaction.

Nucleic Acids

Nucleic acids are complex organic molecules that consist of carbon, hydrogen, oxygen, nitrogen, and phosphorus. There are two major kinds of nucleic acids. Table 2-5 lists some facts about nucleic acids.

Table 2-5 Nucleic Acid Facts	
Deoxyribonucleic Acid	**Ribonucleic Acid**
Abbreviated as DNA	Abbreviated as RNA
Makes up the strands of chromosomes	Found in the nucleolus
Found in the nucleus	Can act as messenger RNA (mRNA)
Gives a coded message to RNA to instruct the ribosomes to make protein	Transfer RNA (tRNA) is found in the cytosol of the cell.

Example Problems

1. True or false: The main characteristic of lipids is the fact that they contain glycerol and fatty acids.

    answer: false. Cholesterol is a lipid and it does not contain glycerol or fatty acids. The main characteristic of lipids is the fact that they are molecules that are insoluble in water.

2. Glucose is a molecule that can enter through the cell membrane via the protein channels. Lactose is an organic molecule found in milk. Lactose is too large to pass through the protein channels, therefore, that we have to digest lactose. When we digest lactose, what two organic molecules are produced?

    answer: glucose and galactose. Lactose is made of glucose and galactose. When it is digested, the digestive enzymes will "break" the bond between the two units, thereby producing individual units of glucose and galactose.

3. Several amino acids bonded together will produce a _____.

    answer: protein. Amino acids are the building blocks of protein.

4. DNA molecules are found in the _____ of a cell while RNA can be
    found in the _____.

    answer: nucleus; nucleolus. DNA is found in the nucleus of the cell and RNA is found in
    the nucleolus, which is a structure within the nucleus.

5. True or false. Cholesterol is classified as a lipid because it is made of the same molecules
    as other lipids.

    answer: false. Cholesterol is a lipid because it is insoluble in water. Cholesterol does not
    consist of the same molecules as other lipids.

pH Concepts

pH is the measure of hydrogen ions in a given solution. The more hydrogen ions present, the
greater the acidity. Hydrogen ions have an acidic characteristic. The pH of a solution is typically
represented on a pH scale that ranges from 1 to 14. A pH of 1 is very acidic and a pH of 7 is
neutral and a pH of 14 is very alkaline (basic) or is the least acidic.

The main concept of understanding pH is to know that each number on the pH scale actually
represents a factor of 10. For example, if a solution is supposed to exhibit a pH of 6 but all of a
sudden the pH drops to 5, it is not a matter of just a single number but is actually 10 times more
acidic. If the pH drops down to 4, it will be 100 times more acidic than pH 6.

Example Problems

For the following problems, determine how many times more acidic or more alkaline one solution
is compared to the other.

1. How many times more acidic is pH 6 compared to pH 8?

    answer: 100; going from pH 8 to pH 7 is 10 times more acidic. Going from pH 7 to pH 6
    is another 10 times more acidic. $10 \times 10 = 100$.

2. How many times more acidic is pH 2 compared to pH 5?

    answer: 1000; going from pH 5 to pH 4 is 10 times more acidic. Going from pH 4 to pH 3
    is another 10 times more acidic. Going from pH 3 to pH 2 is another 10 times. $10 \times 10 \times 10 = 1,000$.

3. How many times more acidic is pH 9 compared to pH 10?

    answer: 10; going from pH 10 to pH 9 is 10 times more acidic.

4. How many times more alkaline is pH 7 compared to pH 2?

    answer: 100,000; going from pH 2 to pH 3 is 10 times more alkaline. Going from pH 3
    to pH 4 is another 10 times more alkaline. Going from pH 5 to pH 6 is another 10 times
    more alkaline. Going from pH 6 to pH 7 is another 10 times more alkaline.

    $10 \times 10 \times 10 \times 10 \times 10 = 100,000$.

5. How many times more alkaline is pH 6 compared to pH 4?

    answer: 100; going from pH 4 to pH 5 is 10 times more alkaline. Going from pH 5 to pH 6 is another 10 times more acidic. $10 \times 10 = 100$.

Buffers

In order for chemical reactions to occur properly in the human body, the enzymes must be functioning properly. In order for the enzymes to function properly, the pH has to be just right. Consider the following areas of the body with their precise pH for normal enzyme function (therefore normal chemical reactions):

❑ Stomach: 1–2

❑ Small intestine: 7–8

❑ Blood: 7.35–7.45

❑ Urine: 5–8

A buffer is a substance that can stabilize the pH. The pH value can change due to the accumulation of hydrogen ions (H^+) or to the lack of hydrogen ions. If the system is beginning to accumulate hydrogen ions, the buffers will bond to the ions, thereby removing them and causing the hydrogen ions to not behave as an acid, thereby stabilizing the pH. If the system is lacking hydrogen ions, buffers can release hydrogen ions into the solution, thereby stabilizing the pH.

Work Problems

1. How many double bonds would be found in this molecule: C   C   O

    O

2. How many triple bonds would be found in this molecule: N   C   H

3. What two organic molecules, when combined, will form sucrose (table sugar)?

4. What are the chemical substances that make up a phospholipid molecule?

5. Protein molecules are made of individual monomers called _____.

6. How many times more acidic is a pH of 5 compared to pH 8?

7. How many times more basic is a pH of 9 compared to pH 7?

8. The pH of blood is slightly (acidic or basic).

9. An accumulation of what ion will cause a system to become more acidic?

10. A substance that causes the pH of a system to remain relatively stable is called a _____.

Worked Solutions

1. $C = C = O$. There will be a total of three double bonds.

    $||$

    O

2. $N \equiv C - H$. There will be one triple bond.

3. Glucose bonded to fructose will yield sucrose.

4. One glycerol, two fatty acids, and one phosphate ion.

5. amino acids

6. 1,000 times more acidic. Going from pH 8 to a pH 7 = 10 times more acidic. Going from pH 7 down to pH 6 = 10 times more acidic. Going from pH 6 down to pH 5 = 10 times more acidic. Therefore, $10 \times 10 \times 10 = 1000$.

7. 100 times more basic. Going from pH 7 to a pH 8 = 10 times more basic. Going from pH 8 to pH 9 = 10 times more basic. Therefore, $10 \times 10 = 100$.

8. basic. The pH of blood is 7.35–7.45. Anything above pH 7 is considered basic.

9. An accumulation of hydrogen ions (H^+) will cause the pH of a system to become more acidic.

10. Anything that resists changes in pH is called a buffer.

Chapter Problems and Solutions

Problems

1. How do you determine the number of neutrons an isotope has?

2. When calcium becomes an ion, it will develop an ionic charge of _____.

3. Bond the following ions together: Sodium ions and nitrogen ions.

4. Bond the following ions together: Magnesium ions and phosphate ions.

5. How many single bonds and double bonds would be found in this organic molecule?

    H   C   O

        H

6. What is the name of the saccharides produced after the digestion of sucrose?

7. Why are fats typically referred to as triglycerides?

8. Which statement is true:

    a. Fats are a type of lipid.

    b. Lipids are a type of fat.

9. If acid begins to accumulate in the blood, the pH will typically stay at about 7.35 to 7.45 because of the presence of what type of organic molecule?

10. A change of one number on the pH scale is actually rather serious because each number on the pH scale represents a factor of _____.

Answers and Solutions

1. **Subtract the number of protons from the AMU value (AMU − protons = neutrons).**

2. **2+.** Calcium is in the second column on the Periodic Table. Elements in the second column develop a 2+ charge when they become an ion.

3. **Na_3N.** Sodium ions have a 1+ charge (Na^{1+}). Nitrogen ions typically have a 3− charge (N^{3-}). Take the 1 (from the 1+ associated with Na) and place it with the N in the subscript form (N_1). Take the 3 (from the 3− associated with N) and place it with the Na in the subscript form (Na_3).

4. **$Mg_3(PO_4)_2$.** Magnesium ions have a 2+ charge (Mg^{2+}). Phosphate ions have a 3− charge (PO_4^{3-}). Take the 2 (from the Mg^{2+}) and place it with the phosphate in the subscript form ($PO_4)_2$. You need to have 2 phosphates. Take the 3 (from the PO_4^{3-}) and place it with the Mg in the subscript form (Mg_3).

5. **There will be two single bonds and one double bond.** H ———— C ══ O
   |
   H

6. **glucose and fructose**

7. **Fats consist of three fatty acids.**

8. **a.** Fats are a type of lipid. Lipid is a category of molecules consisting of fats, phospholipids, glycolipids, and cholesterol.

9. **Buffers are molecules that resist changes in pH.**

10. **Each number represents a factor of 10.**

Supplemental Chapter Problems

Problems

1. When strontium (Sr) becomes an ion, it will develop a charge of _____.

2. When chlorine (Cl) becomes an ion, it will develop a charge of _____.

3. How many electrons would a neutral atom have if it had 22 protons? _____

4. What subatomic particles are found in the nucleus of an atom? _____

5. If an atom has 13 protons, it must be (identify the atom) _____.

6. ^{138}Ba has an AMU of _____.

7. Put in the correct covalent bonds in this molecule: H   C   C   H

8. Glucose bonded to _____ will produce lactose.

9. Which solution is more acidic: a solution with a lot of hydrogen ions or a solution with very few hydrogen ions? _____

10. Chromosomes are strands of material consisting of molecules of _____.

Answers

1. 2+ charge. Look at the Periodic Table, second column.

2. 1– charge. Look at the Periodic Table, seventh major column.

3. 22. The electron number and proton number are the same in a neutral atom.

4. protons and neutrons. Electrons are orbiting the nucleus.

5. Aluminum has 13 protons.

6. 138. The number in the superscript left side is the AMU value for that atom.

7. H ⎯ C ≡ C ⎯ H. Each hydrogen has to have one bond and each carbon has to have a total of four bonds.

8. galactose

9. A solution with the most hydrogen ions is the most acidic.

10. Molecules of deoxyribonucleic acid (DNA) make up the strands of chromosomes.

Chapter 3
The Cell

The human body consists of over 75 trillion cells. In order for the body to maintain homeostasis, the cells must be functioning properly and, therefore, be in homeostasis. In order for the cells to be in homeostasis, the cell organelles must be functioning properly. This chapter discusses the normal function of the various cell organelles and the cell membrane.

The Cell Organelles

Consider the following bullet list, which shows the various cell organelles and structures and gives a brief description of their function.

- ❑ **Nucleus:** Consists of 46 chromosomes, which are made of DNA molecules.

- ❑ **Nucleolus:** Located in the nucleus; produces RNA.

- ❑ **Ribosomes:** Produce protein; can be free or fixed to endoplasmic reticulum.

- ❑ **Golgi apparatus:** Modifies the protein that is produced by ribosomes. Many times, it will add a carbohydrate to the protein, thus producing a glycoprotein. It also produces lysosomes, which consist of enzymes.

- ❑ **Lysosomes:** Vacuoles that contain enzymes. Enzymes will digest organic molecules.

- ❑ **Mitochondria:** Produces 95% of the ATP required for cell activity; produces cholesterol.

- ❑ **Rough endoplasmic reticulum:** Network of channels passing throughout the cytoplasm; have ribosomes attached to them (fixed ribosomes).

- ❑ **Smooth endoplasmic reticulum:** Produces lipids and carbohydrates; do not have ribosomes attached to them.

Protein Synthesis

The ribosomes make protein, but they do not make the protein at random. The ribosomes are "told" what kind of protein to make and when to make it via the direction of DNA. However, DNA is too large of a molecule to exit the nucleus, so there has to be another way to get the instructions out to the ribosomes. That's where RNA comes into play. RNA leaves the nucleolus and picks up the chemical coded message from the DNA. The process of "picking up" the coded message is called **transcription.** RNA then leaves the nucleus with the coded message and travels to the ribosomes. Once at the ribosomes, the RNA (now called mRNA or messenger RNA) gives the message to the ribosomes. The ribosomes take this message and begin to manufacture protein. The process of interpreting the message is called **translation.**

The Cell Membrane

The cell membrane is a semipermeable, bilayered structure made of four major organic molecules, as follows:

- ❑ **Phospholipid:** Creates the semipermeability of the membrane; makes up the majority of the two layers of the membrane. A phospholipid molecule consists of a glycerol molecule bonded to two fatty acids and one phosphate.

- ❑ **Glycolipid:** Gives the cell its identity; located only in the outer layer. A glycolipid molecule consists of a glycerol molecule bonded to two fatty acids and one carbohydrate.

- ❑ **Cholesterol:** Appears to control the rate of flow of material in and out of the cell.

- ❑ **Protein:** Many will form channels to allow larger molecules to enter into the cell.

Example Problems

1. Which cell organelle is responsible for turning a protein molecule into a glycoprotein?

   answer: Golgi apparatus

2. Which cell organelle is located inside the nucleus?

   answer: nucleolus

3. Which cell organelle consists of enzymes that can be used to digest large organic molecules?

   answer: lysosomes

4. Glycolipids are associated with which layer of the cell membrane?

   answer: outer layer

5. Which membrane structure creates the "pores" of a cell?

   answer: protein

Work Problems

Use cell organelle terms to answer Questions 1 through 4.

1. Identify the cell organelle that is involved in making adenosine triphosphate.

2. Identify the cell organelle that is involved in producing lysosomes.

3. Identify the cell organelle that is involved in producing carbohydrates.

4. Smooth endoplasmic reticulum is called "smooth" because it lacks _____.

Use cell membrane terminology to answer Questions 5 through 7.

5. The cell membrane structure that is located only in the outer layer is the
   _____ molecule.

6. How many layers make up the cell membrane?

7. Large molecules, such as glucose, cannot pass easily through the cell membrane. Therefore, those molecules must pass through the _____ channels.

Worked Solutions

1. **The mitochondria make ATP.**

2. **The Golgi apparatus makes lysosomes.**

3. **The smooth endoplasmic reticulum produces carbohydrates and lipids.**

4. **Smooth endoplasmic reticulum lacks ribosomes.**

5. **The glycolipid molecule is located only on the outer layer.**

6. **There are 2 layers that make up the single membrane.**

7. **Protein channels are formed that will allow large molecules such as glucose to enter the cell.**

Osmosis

In order to understand osmosis, you must be familiar with the concept of diffusion. **Diffusion** is defined as the movement of **molecules** from an area of high concentration to an area of low concentration. By comparison, **osmosis** is defined as the movement of **water molecules** from an area of high concentration to an area of low concentration **across a membrane.** Osmosis is a special type of diffusion, because it involves the movement specifically of water molecules. Examples of diffusion would be:

❑ The spreading of perfume throughout the room

❑ The smell of armpit odor from a person with hygiene problems

❑ Osmosis

In order to understand osmosis, you must understand the following terms:

❑ **Solution** is a term that refers to a combination of the solute and solvent.

❑ **Solute** is a term that refers to the material that is dissolved in the solvent.

❑ **Solvent** is a term that refers to the liquid that is doing the dissolving. In reference to osmosis, the solvent is water and the solute is typically an ion, saccharide, or mineral.

When studying osmosis, pretend that the only items that comprise a cell are water and solutes. The inside of the cell is known as the **ICF (intracellular fluid),** and the outside of the cell is known as the **ECF (extracellular fluid).**

The best way to study and understand the concept of osmosis is to examine and use the following four steps. In all cases, we will allow the water to move but have the solutes remain stationary.

1. Determine which (the ICF or the ECF) has the highest concentration of water.

2. Determine whether water will flow into the ICF or into the ECF.

3. Determine which (the ICF or the ECF) has the highest concentration of solutes.

4. Determine an osmotic term for the ICF and the ECF.

The key osmotic terms are:

❑ **Hypotonic:** The ICF is hypotonic to the ECF if it consists of fewer solutes than the ECF.

❑ **Hypertonic:** The ICF is hypertonic to the ECF if it consists of more solutes than the ECF.

❑ **Isotonic:** The ICF is isotonic to the ECF if it has the same concentration of solutes as the ECF.

Examine the osmotic examples in Figure 3-1:

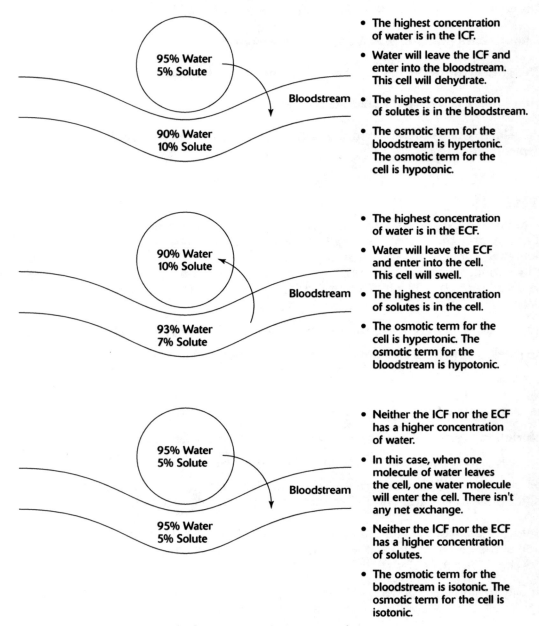

Figure 3-1: Osmosis examples.

Keep the following other notes in mind:

❑ Water always flows from the hypotonic area to the hypertonic area.

❑ The osmotic terms always refer to the **solute** concentrations in the ICF and the ECF.

❑ Whenever the ICF is hypotonic, the ECF is automatically hypertonic and vice-versa.

❑ When the solute concentration in an area increases, the water concentration automatically goes down. The combination of the solute percentage and the water percentage equals 100.

❑ When the water concentration in an area increases, the solute concentration automatically goes down. The combination of the solute percentage and the water percentage equals 100.

Example Problems

For the following problems, determine whether water will enter the cell or exit the cell. Then, determine the osmotic term for the ICF and the ECF.

1.

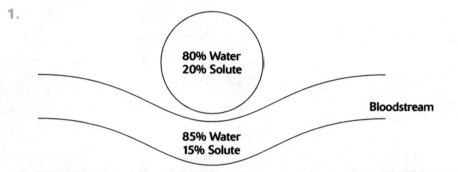

answer: Water will enter the cell. The cell is hypertonic and the bloodstream is hypotonic.

2.

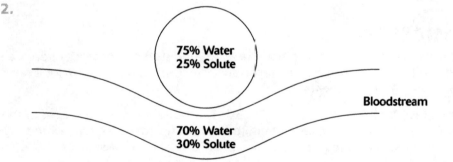

answer: Water will leave the cell. The cell is hypotonic and the bloodstream is hypertonic.

3.

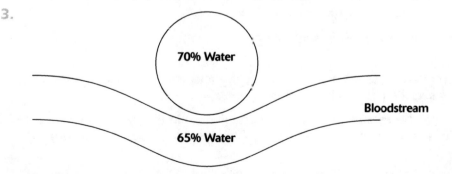

answer: Water will leave the cell. In this case, you can calculate the percentage of solutes in the cell and the bloodstream. The cell consists of 30% solutes and the bloodstream consists of 35% solutes. Therefore, the cell is hypotonic and the bloodstream is hypertonic.

4.

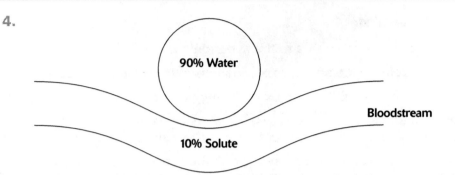

answer: In this case, you need to calculate the percentage of solutes in the cell and calculate the percentage of water in the bloodstream. After doing that, you will find that there are equal amounts of water in the cell and the bloodstream. Because the amounts of solutes are the same, the cell is isotonic and the bloodstream is also isotonic.

5.

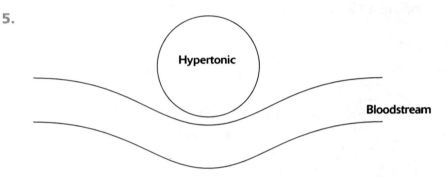

answer: In this case, because the cell is hypertonic, the bloodstream is automatically hypotonic. Because the bloodstream is hypotonic, it therefore has fewer solutes than the cell and therefore more water than the cell. Water will enter into the cell.

Cell Reproduction

In order to repair wounds, the various cells of the body need to undergo cell reproduction. Cell reproduction is a very controlled set of events. Cell reproduction consists of three major events. They are: **interphase, mitosis,** and **cytokinesis.**

Mitosis consists of several very characteristic phases. Mitosis is a set of phases that occur in the nucleus of the cell. The various mitosis phases are identified according to the activity of the chromosomes inside the nucleus. The mitosis phases are: **prophase, metaphase, anaphase,** and **telophase.**

Therefore, cell reproduction consists of all of the following phases: interphase, prophase, metaphase, anaphase, telophase, and cytokinesis:

- ❏ **Interphase:** Everything inside the cell is duplicating; you cannot see the chromosomes inside the nucleus.

- ❏ **Prophase:** The chromosomes have completed doubling and are now called paired chromatids; you can now see the paired chromatids.

- ❏ **Metaphase:** The paired chromatids move around and line up in the middle of the nuclear region. The centrioles produce spindle fibers, which connect to each paired chromatid.

❏ **Anaphase:** The paired chromatids are separated. The spindle fibers retracts, thus separating the paired chromatids.

❏ **Telophase:** The nuclear region pinches in half to create two nuclei.

❏ **Cytokinesis:** The cell membrane pinches in half to create two cells.

Example Problems

The Figure 3-2 shows the various phases of cell reproduction out of order. After responding to the questions, you will be able to put the figures in correct sequence. The figure that represents interphase shows two chromosomes. Technically, you cannot see the chromosomes. The centrioles and spindle fibers are not shown in the figure.

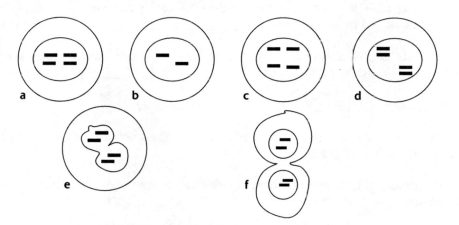

Figure 3-2: Cell reproduction examples.

1. Which part of the figure shows the paired chromatids separating?

   answer: c

2. Which part of the figure shows the nuclear region pinching in half?

   answer: e

3. Which part of the figure shows the paired chromatids lining up in the middle of the nuclear region?

   answer: a

4. Which two parts of the figure represent interphase and cytokinesis?

   answer: b represents interphase and f represents cytokinesis.

5. Place the parts of the figures in the correct sequence.

   answer: b, d, a, c, e, f

Work Problems

Use osmotic terms to describe the following scenarios in Questions 1 through 5.

1. If a cell consists of 25% solutes and the solution the cell is exposed to consists of 30% solutes, the cell would be considered _____ to the solution.

2. If a cell consists of 10% solutes and the solution the cell is exposed to consists of 5% solutes, the cell would be considered _____ to the solution.

3. If a cell consists of 8% solute and the solution the cell is exposed to consists of 80% water, the cell would be considered _____ to the solution.

4. If a cell consists of 90% water and the solution the cell is exposed to consists of 80% water, the cell would be considered _____ to the solution.

5. If a cell consists of 20% solute and the solution it is exposed to also consists of 20% solute, the cell would be considered _____ to the solution.

For Questions 6 through 10, use cell reproduction terminology.

6. Identify the phase of cell reproduction where the paired chromatids are lining up in the middle of the nuclear region.

7. Identify the phase of cell reproduction where the nuclear region is pinching in two thereby resulting in two new nuclei.

8. Identify the phase of cell reproduction where the cell membrane is pinching in two thereby resulting in two new cells.

9. Identify the first phase of mitosis.

10. Identify the first phase of the cell life cycle.

Worked Solutions

1. **hypotonic.** Because the cell consists of fewer solutes than the solution it is exposed to, it is considered hypotonic.

2. **hypertonic.** Because the cell consists of more solutes than the solution it is exposed to, it is considered hypertonic.

3. **hypotonic.** Because the solution the cell is exposed to consists of 80% water, it therefore consists of 20% solutes. The cell therefore consists of fewer solutes. Having fewer solutes is hypotonic.

4. **hypotonic.** Because the cell consists of 90% water, it therefore consists of 10% solute. Because the solution the cell sits in is 80% water, it therefore consists of 20% solute. The cell therefore consists of fewer solutes than the surrounding solution. Having fewer solutes is hypotonic.

5. **isotonic.** Because the cell consists of the same concentration of solutes as the solution it sits in, the cell is considered to be isotonic.

6.   **metaphase**

7.   **telophase**

8.   **cytokinesis**

9.   **prophase**

10.   **interphase**

Chapter Problems and Solutions

Problems

1.   Water molecules are small enough to pass through the phospholipid layers of the cell membrane but larger molecules such as glucose have to pass through a specific portion of the cell. What is that structure called?

2.   Identify the phase of cell reproduction where chromatids are forming.

3.   Identify the phase of cell reproduction where chromatids can now be seen.

4.   There are some chemical reactions occurring in the cytosol (cytoplasm) of the cell that will produce about 5% of the cell's energy in the form of ATP. What cell structure produces 95% of the cell's energy?

5.   If a cell is losing water to its environment (ECF) the ECF must therefore be (hypotonic or hypertonic)?

6.   If a cell is gaining water from the ECF, the ECF must therefore be (hypotonic or hypertonic)?

7.   If a cell is hypotonic, the ECF must be _____.

8.   The osmotic terms are always terms that refer to (water or solutes).

9.   If a cell suddenly has an increase in solutes, the _____ concentration will automatically decrease.

10.   An isotonic cell contains the same concentration of _____ as the ECF.

Answers and Solutions

1.   **Larger molecules like glucose pass through protein channels to enter into the cell.**

2.   **interphase.** You can see the paired chromatids in prophase, but they were formed during interphase.

3.   **prophase.** Everything is in the process of doubling in interphase but cannot be seen until prophase.

4. **mitochondria**

5. **hypertonic.** The ECF must be hypertonic, which means it has more solutes and less water than the cell. Because the ECF has less water than the cell, water will leave the cell.

6. **hypotonic.** The ECF must be hypotonic, which means it has fewer solutes and more water than the cell. Because the ECF has more water than the cell, water will enter the cell.

7. **hypertonic.** Because the cell is hypotonic, the cell has fewer solutes. It therefore has fewer solutes than the ECF. Therefore, the ECF has more solutes and is therefore considered to be hypertonic.

8. **The osmotic terms are only solute terms.**

9. **The total percentage of solutes and water must equal 100%.** If the solute concentration goes up, the water concentration goes down to maintain a balance of 100%.

10. **solute.** The correct explanation would use the word solute, because the osmotic terms always refer to solute values and not water values.

Supplemental Chapter Problems

Problems

1. What happens to the chromatids during anaphase?

2. The ultimate formation of two new nuclei occur during which phase of cell reproduction?

3. What are the four main molecular components of a cell membrane?

4. An ECF that has a higher concentration of solutes than the ICF would be considered what osmotic term?

5. What osmotic term describes the ECF in this scenario? A cell is exposed to an ECF that does not result in the gain or loss of any water.

6. The major function of mitochondria is to produce energy that is in the form of _____.

7. A cell is beginning to swell after it was exposed to a solution. The solution (ECF) must therefore be _____ (use an osmotic term).

8. Which cell organelle synthesizes carbohydrates?

9. Which complex molecule (DNA or RNA) actually delivers a coded message to the ribosomes for the purpose of manufacturing protein?

10. The process by which the DNA gives the coded message to RNA is called _____.

11. DNA makes up the chromosomes, which are located in the nucleus of the cell and RNA is located in the _____, which is inside the nucleus of the cell.

12. During which phase of cell reproduction are the chromosomes in the process of duplicating?

13. Chemotherapy drugs will kill cancer cells by inhibiting the separation of the chromatids. If the chromatids cannot separate, the cell will die because it cannot finish the process of cell reproduction. Based on this, what phase does chemotherapy drugs have an affect on?

14. If something happens to the DNA (mutation for example), then the DNA will give the RNA an abnormal message. This could ultimately cause the ribosomes to make an abnormal _____.

15. Streptomycin is a drug that kills bacteria. Streptomycin causes the bacterial ribosomes to malfunction. Therefore, the bacteria cannot produce _____, which they need for survival.

Answers

1. The chromatids are separated and will move to opposite ends of the nuclear region.

2. telophase. Two new cells are formed during cytokinesis.

3. phospholipids, glycolipids, cholesterol, and protein

4. The ECF would be hypertonic to the ICF.

5. The ECF would be isotonic to the ICF (cell).

6. ATP (adenosine triphosphate)

7. hypotonic

8. smooth endoplasmic reticulum

9. RNA (ribonucleic acid). When it delivers the coded message, it is called messenger RNA (mRNA).

10. transcription

11. nucleolus

12. interphase. You can see them in prophase but the doubling occurs in interphase.

13. anaphase

14. protein

15. protein

Chapter 4
Cells and Tissues

The human body consists of trillions of cells, but they don't work independently. The cells work together to perform various tasks to keep the body in homeostasis. In order for the cells to work together, they form tissues.

Tissues do not work independently, either. The tissues work together to perform various tasks to keep the body in homeostasis. In order for the tissues to work together, they form organs.

In order to study the various cells and tissues that comprise the body, anatomists have subdivided the tissues into four major groups and those groups are further subdivided according to their cellular makeup.

The Four Tissue Groups

Table 4-1 lists the four major tissue groups and a characteristic for each group. These major tissue groups are discussed in detail in the following sections.

Table 4-1 Tissue Types		
Tissue Type	**Characteristic**	**Type of Cell Within the Tissue Type**
Epithelial	This tissue type consists of cells that make up the inside or the outside lining of organs.	Squamous, cuboidal, columnar
Muscular	This tissue type consists of cells that have the ability to contract and relax.	Skeletal muscle, smooth muscle, cardiac muscle
Neural	This tissue type consists of cells that conduct impulses or cells that protect the nervous system.	Neuron, glial
Connective	This tissue type consists of cells that have a matrix and typically fill internal spaces within the body.	Adipose, areolar, blood, bone, cartilage, dense, reticular

In order to fully understand how the cells and tissues are involved in helping the body maintain homeostasis, you need to understand a few basic facts about each type of cell and tissue.

Epithelial

Squamous cells make up epithelial tissue:

- ❑ Squamous cells are flat in appearance (refer to Figure 4-1).
- ❑ These cells can be found lining the skin.
- ❑ These cells are our first line of defense.

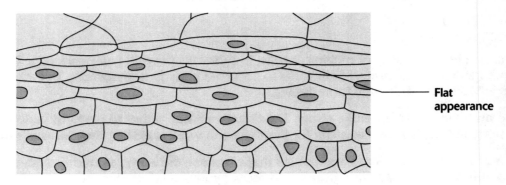

Flat
appearance

Figure 4-1: Squamous cells.

Cuboidal cells make up epithelial tissue:

- ❑ Cuboidal cells are shaped like little squares (refer to Figure 4-2).
- ❑ These cells can be found lining the urinary tubes.
- ❑ These cells secrete and absorb material.

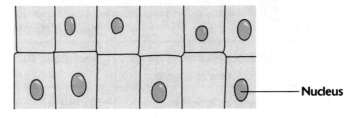

Nucleus

Figure 4-2: Cuboid cells.

Columnar cells make up epithelial tissue:

- ❑ Columnar cells are shaped like columns (refer to Figure 4-3).
- ❑ These cells can be found lining the trachea.
- ❑ These cells secrete and absorb material.

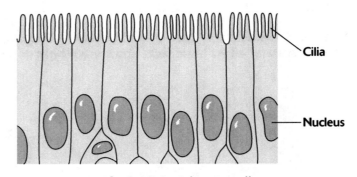

Cilia

Nucleus

Figure 4-3: Columnar cells.

Muscular

Skeletal muscle cells make up muscular tissue:

❏ Skeletal muscles are elongated cells with striations (refer to Figure 4-4).

❏ These cells can be found making up the muscles associated with the skeletal system.

❏ These cells contract and relax under voluntary control.

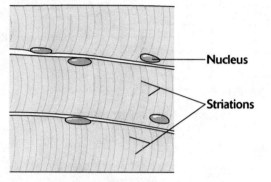

Figure 4-4: Skeletal muscle cells.

Smooth muscle cells make up muscular tissue:

❏ Smooth muscles are elongated cells without striations (refer to Figure 4-5).

❏ These cells can be found making up the uterus and blood vessels.

❏ These cells contract and relax under involuntary control.

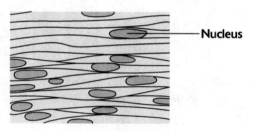

Figure 4-5: Smooth muscle cells.

Cardiac muscle cells make up muscular tissue:

❏ Cardiac muscle cells consist of intercalated disks (refer to Figure 4-6).

❏ These cells can be found making up only the heart.

❏ These cells contract and relax in a pulsating manner.

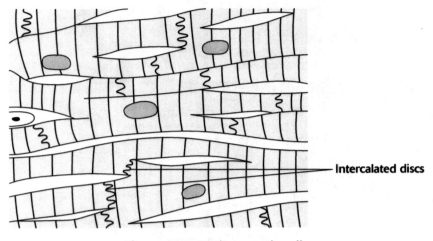

Figure 4-6: Cardiac muscle cells.

Neural

Neurons make up neural tissue:

❑ Neurons consist of dendrites, soma, and an axon (refer to Figure 4-7).

❑ These cells can be found making up the nervous system such as the brain and spinal cord.

❑ These cells conduct impulses.

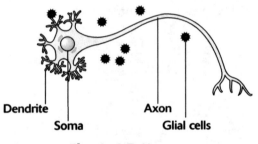

Figure 4-7: Neuron.

Glial cells make up neural tissue:

❑ Glial cells have a variety of sizes and shapes.

❑ These cells can be found either nearby or surrounding the neurons (refer to Figure 4-7).

❑ These cells provide protection for the neurons.

Connective

Connective tissue represents the most diverse tissue group of the human body. Even though the cells are quite different from each other, they do have some commonality. The main commonality between them is the fact that they all have a matrix of some sort. The matrix is the material that surrounds the cells. There are four major types of matrix material.

❑ **Fiber:** The fibers of the matrix can be long, slender fibers or short, thick fibers.

❑ **Liquid:** The matrix can be liquid such as the plasma of blood.

❑ **Solid:** The matrix can be solid, which creates a tough, strong type of tissue.

❑ **Gel:** The gel matrix creates a tough but yet very flexible type of tissue.

Adipose cells make up connective tissue:

❑ Adipose cells are round and appear to be empty, but they're actually full of molecules of fat (see Figure 4-8).

❑ These cells can be found surrounding various organs of the body.

❑ These cells provide insulation.

❑ These cells have a fiber matrix.

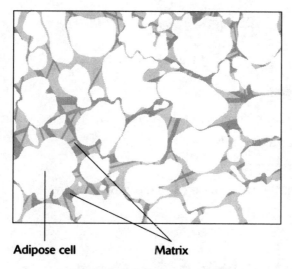

Adipose cell   **Matrix**

Figure 4-8: Adipose cells.

Areolar cells make up connective tissue:

❑ Areolar cells are small and have long, thin fibers running between them making up the matrix (refer to Figure 4-9).

❑ These cells can be found between our skin and muscle.

❑ These cells provide attachment of our skin to the muscle.

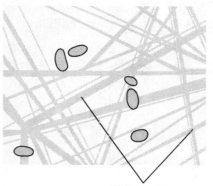

Fiber matrix

Figure 4-9: Areolar cells.

Blood cells make up connective tissue:

- ❏ Blood cells are small anucleated (do not have a nucleus) cells with a plasma matrix (refer to Figure 4-10).
- ❏ These cells can be found in our circulatory system.
- ❏ These cells transport oxygen and carbon dioxide to and from other cells of the body.
- ❏ These cells have a liquid matrix that we call plasma.

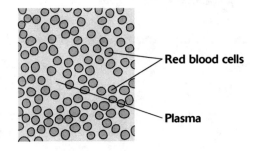

Figure 4-10: Blood cells.

Bone cells make up connective tissue:

- ❏ Bone cells form concentric rings around a central canal (refer to Figure 4-11).
- ❏ These cells can be found in our bones.
- ❏ These cells provide strength.
- ❏ These cells have a solid matrix made of calcium phosphate.

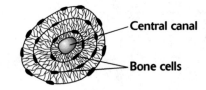

Figure 4-11: Bone cells.

Cartilage cells make up connective tissue:

- ❏ Cartilage cells are small and sit in a rather large lacuna (see Figure 4-12).
- ❏ These cells can be found within our joints.
- ❏ These cells reduce friction in the joints.
- ❏ These cells have a gel matrix, which allows for flexibility.

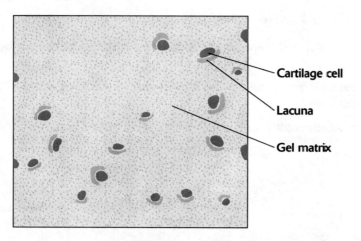

Figure 4-12: Cartilage cells.

Dense cells make up connective tissue:

❑ Dense cells are parallel fibers (refer to Figure 4-13).

❑ These cells make up tendons and ligaments.

❑ Tendons provide attachment of muscles to bone while ligaments provide attachment of one bone to another.

❑ These cells have a fiber matrix.

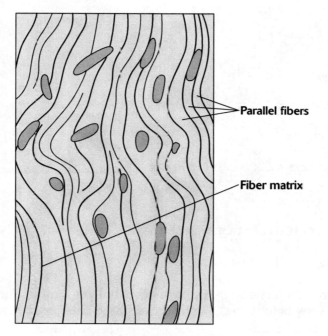

Figure 4-13: Dense cells.

Reticular cells make up connective tissue:

❑ Reticular cells are small and have short, thick fibers running between them (refer to Figure 4-14).

❑ These cells can be found making up the framework of the liver, spleen, tonsils, appendix, and thymus gland.

❑ These cells have a fiber matrix that consists of short, thick fibers.

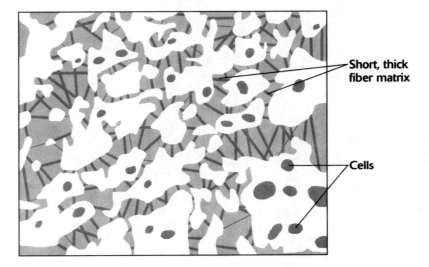

Short, thick fiber matrix

Cells

Figure 4-14: Reticular cells.

Example Problems

1. Which type of cells have intercalated discs?

   answer: cardiac cells

2. Cells that contract and relax are found in the _____ tissue category.

   answer: muscular

3. What is the name of the cells that typically have a gel matrix?

   answer: cartilage

4. Blood cells do not connect one tissue to another but yet they still belong in the connective tissue category. They belong in the connective tissue category because they have a

   _____.

   answer: matrix

5. The liver consists of cells that have what type of matrix?

   answer: short, thick fiber matrix

Body Membranes

There are four major types of membranes in the body. They are:

❑ **Cutaneous membrane:** The cutaneous membrane is actually the skin. It covers the entire body and is made of squamous epithelial cells.

❑ **Mucous membranes:** The mucous membranes typically consist of columnar epithelial cells. These membranes produce mucus, which provides protection to the tissue they line. The mucous membranes line the cavities that open to the outside of the body, such as:

The digestive tract: Opens to the outside via the mouth and anus.

The respiratory tract: Opens to the outside via the mouth and nose.

The reproductive tract: Opens to the outside via the vaginal opening.

The urinary tract: Opens to the outside via the urethral opening.

❑ **Serous membranes:** The serous membranes produce serous fluid, which provides protection for the tissue they cover. These membranes are made of epithelial cells that are supported by connective tissue. The serous membranes line the internal cavities of the body that are not open to the outside, such as:

The **pleural membranes** cover the lungs.

The **peritoneal membranes** cover organs such as the stomach and liver.

The **pericardial membrane** covers the heart.

❑ **Synovial membranes:** The synovial membranes produce synovial fluid. This fluid helps to reduce abrasion at the joint site. These membranes are made of connective tissue. The synovial membranes line the joints of the body.

Example Problems

1. Which type of membrane is made mostly of connective tissue?

   answer: synovial membrane. Synovial membrane consists of complete connective tissue whereas serous membranes have incomplete connective tissue for support.

2. Which type of membrane is made mostly of squamous cells?

   answer: cutaneous membrane

3. Membranes are made mostly of what two types of tissue?

   answer: epithelial and connective tissue

4. Which type of membrane can be found in the elbow region?

   answer: synovial membrane. Synovial membranes line the joints of the body.

5. Which type of membrane lines the inside of the nasal cavity?

   answer: mucous membrane. Mucous membranes line the cavities that are open to the outside.

Work Problems

1. A collection of cells that work together to perform specific functions is called a
    _____.

2. Which type of cells are typically the first ones to be exposed to environmental hazards?

3. There are cells that line some of the glands of the body and are shaped like little cubes.
    What type of cells are they?

4. Cells that store molecules of fat are called _____ cells.

5. Chondrocytes are cells that have a gel matrix. Therefore, chondrocytes are what type of cell?

6. The reduction of friction between the two surfaces within the internal cavity of the body is
    the function of what type of membrane?

7. Which type of muscle cell has the ability to contract rhythmically?

8. The watery substance of blood is a matrix called _____.

9. Pretend you discovered a new type of cell in the body. Upon observation, you observed
    some small fibers between the individual cells. What tissue category would these new
    cells belong to?

10. What type of tissue makes up the inside lining of the mouth?

Worked Solutions

1. **tissue**

2. **squamous cells.** These cells make up our skin. The skin is the first line of defense.

3. **cuboid cells.** Cuboid cells are shaped like squares or little cubes.

4. **adipose**

5. **cartilage.** Cartilage consists of a gel matrix.

6. **serous membrane.** The serous membranes are not exposed to the outside of the body.

7. **cardiac muscle cells**

8. **plasma**

9. **Because these cells are associated with a fiber matrix, they would be placed in the
    connective tissue category.** Connective tissue has matrix.

10. **Epithelial tissue makes up the inside lining of the mouth.** Epithelial tissue makes up the
    inside or outside lining of something.

Chapter Problems and Solutions

Problems

1. What type of membrane would line the ear canal?

2. Which type of membrane is made of cells that appear flat?

3. Which type of membrane provides the body with the first line of defense?

4. If you discovered new cells that appeared to line the outside of a piece of tissue, it would probably belong in which tissue category?

5. Most cells of the body can be found in several places. However, there is one type of cell that is limited to the location of just one organ of the body. What type of cell is this?

6. Blood vessels can dilate and constrict on their own (not under our control). They can do this because they are made of what type of muscle cell?

7. What kind of substance would columnar cells produce?

8. Lamella is a type of matrix that is solid. Which of the cell types consist of lamellae?

9. Which of the cell types do not consist of a nucleus?

10. Once in awhile, a person might have a "runny" nose. Their nose "runs" because the cells that make up the _____ membrane produce mucus.

Answers and Solutions

1. **The mucous membranes are a type of membrane that lines areas that are exposed to the outside of the body.**

2. **The cutaneous membrane is made of squamous cells and squamous cells appear flat.**

3. **The cutaneous membrane covers the lining of the skin.** Many organisms that enter the body have to pass through the skin first before getting to their destination.

4. **It would belong in the epithelial tissue category because it makes up the lining of something.**

5. **Cardiac cells are found only in the heart.**

6. **Smooth muscle cells are muscles that are not controlled by us.** These are involuntary muscle cells. The dilation and constriction of blood vessels is an involuntary response.

7. **Columnar cells are found in the trachea making up the mucous membrane.** The mucous membrane secretes mucus. Therefore, it is the columnar cells that secrete a mucus substance.

8. **Bone cells have a solid matrix.** The matrix of bone is called lamellae.

9. **Red blood cells (erythrocytes) do not have a nucleus.** They are anucleated.

10. **The mucous membrane consists of cells that produce mucus.** Mucus is the "runny" liquid that emerges from the nose.

Supplemental Chapter Problems

Problems

1. Which type of membrane lines the joints of the body?

2. Which type of membrane lines the surfaces of areas that are exposed to the atmosphere?

3. The inside lining of the abdomen is lined with which type of membrane?

4. The membrane that provides us with our first line of defense (prevents infectious agents from entering the body) would be which type of membrane?

5. The pleural membranes that line the lungs are what type of membrane?

6. Identify the type of cell that consists of molecules of fat.

7. Identify the type of cell that reduces friction within the joints.

8. Identify the type of cell that lines the trachea.

9. Labor contractions are not under the patient's control. This is because the uterus is made of what type of muscle cells?

10. Blood is in the connective tissue category because the blood cells are associated with a matrix. What is the name of that matrix?

Answers

1. Synovial membranes line the joint cavity.

2. Mucous membranes line the areas that are open to the outside of the body.

3. Serous membranes line the internal body cavities.

4. Cutaneous membrane is actually the skin. The skin is the first line of defense. Our blood (white blood cells) is the second line of defense.

5. Serous membranes line the lungs. Therefore, the pleural membranes are a type of serous membrane.

6. adipose

7. cartilage

8. columnar

9. Smooth muscle cells are under involuntary control.

10. plasma

Chapter 5
The Integumentary System

The integumentary system, which most people think of as just skin, actually consists of nails, hair, glands, *and* skin. The integument is the most visible part of the body, but is taken for granted by most people (except, perhaps, a dermatologist).

The Skin

An organ is considered to be a structure that is made of two or more tissues. The skin consists of numerous tissue types, and because its surface area is so large, it is considered to be the largest organ of the body. The skin is the cutaneous membrane that is made of two layers; the **epidermis** and **dermis.** Those two layers are made of several sub layers. Check out the following list for facts about these layers:

❑ **Epidermis:** The epidermis is the most outer layer of the skin; it consists of four major individual layers.

Stratum corneum: This is the outermost layer of the epidermis; it consists of squamous cells.

Stratum granulosum: This layer of cells produces a protein substance called **keratin.**

Stratum spinosum: This layer of cells produces a substance called **desmosomes.** Desmosomes connect one layer of tissue to another layer.

Stratum germinativum (basale): This is the deepest layer of the epidermis; this layer consists of cells that can reproduce and therefore repair wounds. This layer also consists of cells that produce **melanin,** which is a pigment that gives the skin its color.

❑ **Dermis:** The dermis consists of two layers of tissue.

Papillary: This layer consists of hair follicles and sebaceous glands. This layer also creates the ridges making up fingerprints.

Reticular: This layer consists of sweat glands.

Deep to the dermis is a layer called the **hypodermis.** This layer is not a part of skin but is many times discussed with the topic of skin. This layer consists of adipose cells and major blood vessels. Connecting the hypodermis to the skeletal muscles is a group of cells called areolar. This is very loose material and is not very strong.

Example Problems

1.   The stratum spinosum produces spine-like substances that literally connect one layer of cells to another. These spine-like substances are called _____

     answer: desmosomes

2.   The epidermis and dermis are collectively called the _____ membrane.

     answer: cutaneous

3.   It takes approximately two weeks for the stratum corneum to be replaced by new skin cells. These new skin cells are derived from which layer?

     answer: stratum germinativum

4.   In order to protect our skin from ultraviolet rays of the sun, our skin produces a pigment to make the skin darker, such as the case of a tan. Which layer consists of cells that produce a pigment to create a darker skin color?

     answer: stratum germinativum. This layer consists of melanocytes.

5.   When a person exercises, they have a tendency to create moisture in the axillary region. Which layer of skin produces this excess moisture?

     answer: reticular layer

Hair

Hair protrudes through the skin by first developing in the hair follicles located in the dermis. Hair provides minimal protection, but the hair on our scalp protects us from ultraviolet rays of the sun. If an insect walks on our arm, the insect will cause a hair to move. When the hair moves, a nerve is activated and is then detected by the body.

In the dermis, there is a smooth muscle (**arrector pili muscle**) that is connected to the shaft of the hair and also to the superficial area of the dermis. There are a variety of stimuli that will cause the arrector pili muscles to contract. Each time the arrector pili muscles contract, the hair will stand straight. When it stands straight, the tissue near the hair will begin to bulge, thus creating a goose bump.

Many times, when a person is chilly, he generates goose bumps. The contraction of thousands of arrector pili muscles generates heat. The shivering of the body is a series of muscle contractions which, upon contraction, will generate heat.

Nails

Fingernails and toenails are made of a tough protein substance called **keratin.** The nails generally appear pinkish in color, which is due to the underlying blood vessels. At the root of the nail, the blood vessels are hidden by running deeper into the tissue. Because of this, the area appears lighter in color. It takes on a half-moon shape and is called the **lunula.** Nails help to protect the fingers and toes.

Glands

Consider the following bullet list of the various glands and their function associated with the integumentary system.

- ❑ **Sebaceous gland:**

    These glands are located in the papillary layer of the dermis.

    These glands produce an oil substance called **sebum.**

    Sebum goes to the surface of the skin and therefore lubricates the skin.

    If the duct of these glands gets blocked, a pimple may occur.

- ❑ **Apocrine gland:**

    These glands are located in the reticular layer of the dermis.

    These glands are a type of sweat gland.

    This type of sweat substance creates the "natural" body odor.

- ❑ **Merocrine gland:**

    These glands are located in the reticular layer of the dermis.

    These glands are a type of sweat gland.

    This type of sweat substance cools the body when it is hot.

- ❑ **Ceruminous gland:**

    These glands are located in the ear canal.

    These glands produce **cerumin** (ear wax).

    Cerumin serves to protect the eardrum.

Example Problems

1. What is the name of the muscle that causes hair to stand erect?

    answer: arrector pili

2. What is the name of the area on the fingernail that is light in color and is located at the root of the nail?

    answer: lunula

3. When perspiration evaporates from the body, the body becomes cooler. Which type of sweat gland produces perspiration?

    answer: merocrine

4. Which gland is involved in acne production?

    answer: sebaceous

5. Which gland produces natural body odor?

    answer: apocrine

Work Problems

1. What are the four main components of the integumentary system?

2. Which layer of the epidermis consists of cells that are undergoing interphase, prophase, metaphase, anaphase, telophase, and cytokinesis?

3. A hypodermic needle is used to place medication nearby some major blood vessels in the skin. Which layer is the hypodermic needle named after?

4. Which glands produce an oily substance?

5. Which gland produces a substance that is designed to keep the skin lubricated and moist, especially when the skin is exposed to a dry environment?

6. There are times when a person has "sweaty" palms. This is because there is a high concentration of a specific type of sweat gland located on the palms. What is the name of that gland?

7. What is the name of the gland located in the ear canal?

8. If a person steps on a thorn, the thorn will penetrate several layers of tissue. If this thorn penetrated to the reticular layer, how many layers of tissue did the thorn penetrate?

9. In order to cool the body, the sweat molecules produced by the _____ glands must evaporate from the surface of the skin.

10. Infants can recognize mom's body odor due to the fact that the nipple regions consists of a high concentration of _____ glands. The infant can smell the secretions each time it nurses.

Worked Solutions

1. **skin, hair, nails, and glands**

2. **stratum germinativum (stratum basale)**

3. **hypodermis layer**

4. **Sebaceous glands produce sebum.** Sebum is oily.

5. **sebaceous glands**

6. **merocrine glands**

7. **ceruminous glands**

8. **Five layers.** The thorn would penetrate the stratum corneum, stratum granulosum, stratum spinosum, stratum germinativum, and the papillary layer. The thorn would penetrate 4 layers of the epidermis and 1 layer of the dermis. There are some parts of the epidermis that have another layer (stratum lucidum) located between the s. corneum and the s. granulosum. If this is the case, the answer is 6 layers.

9. **merocrine**

10. **apocrine**

Chapter Problems and Solutions

Problems

1. Melanocytes are cells that produce melanin. Melanocytes are found in which layer of the skin?

2. An injection given to a patient by a hypodermic needle will inject medication into which layer of the skin?

3. Which layer (epidermis, dermis, or hypodermis) could also be referred to as the subcutaneous?

4. Which gland produces the type of sweat that people, who exercise vigorously, generate?

5. Oil is a sticky substance. Dirt has a tendency to stick to oil. When this happens, dirt will block the release of oil and begin the formation of a pimple. Which gland is being described?

6. The lunula is located at the proximal portion or distal portion of the fingernail?

7. The contraction of lots of _____ muscles could generate a small amount of heat, thereby trying to warm the body.

8. Arrector pili muscles are under (voluntary or involuntary) _____ control.

9. The most superficial layer of the epidermis is the _____.

10. The most variety of glands of the skin are located in which layer?

Answers and Solutions

1. **Melanin is the pigment that is produced by cells in the stratum germinativum layer.** Therefore, melanocytes are found in the stratum germinativum layer.

2. **A hypodermic needle is designed to go deep into the skin into the hypodermis layer, which will, therefore, put the medication near the larger blood vessels.**

3. **Because the epidermis and dermis are collectively called the cutaneous, the layer under the cutaneous would be the subcutaneous (i.e., the hypodermis).**

4. **The merocrine glands produce the type of sweat that is generated when a person exercises, for cooling purposes.**

5. **The sebaceous glands produce sebum, which is oily.**

6. **The lunula is located at the proximal portion of the fingernail, as this is the area that is nearest the main portion of the finger.**

7. **Contraction of the arrector pili muscles could generate a small amount of heat.**

8. **Arrector pili muscles are smooth muscles and are therefore under involuntary control.**

9. The stratum corneum is the most superficial layer of the epidermis.

10. The reticular layer of the dermis consists of two kinds of sweat glands, whereas the papillary layer consists of only the sebaceous glands.

Supplemental Chapter Problems

Problems

1. Which layer of the dermis consists of the sebaceous glands?

2. What is the name of the layer of tissue that is located between the dermis and muscle?

3. What is the name of the most superficial layer of skin?

4. Most of a person's body fat is located in the _____ layer.

5. Melanocytes are cells that produce a pigment that gives the skin its natural color tone. These cells are located in the deepest epidermal layer. What is the name of this layer?

6. Babies have a certain "baby" smell associated with them. This is due to a gland that is very active at a young age. What is the name of that gland?

7. Dandruff is due to rather large "sheets" of epidermal cells flaking off the body. These cells come from the most superficial layer of the epidermis. What is the name of that layer?

8. Which layer of skin consists of the accessory structures such as glands?

9. A blocked sebaceous gland could result in what type of skin condition?

10. Which layer of the dermis consists of the sweat glands?

Answers

1. Sebaceous glands are in the papillary layer of the dermis.

2. The hypodermis is found between the dermis and muscle. Hypodermis is a term that means "below" the dermis.

3. stratum corneum

4. hypodermis

5. stratum germinativum

6. apocrine

7. stratum corneum

8. papillary and reticular layers of the dermis

9. acne or pimples

10. reticular layer

Chapter 6
The Axial Skeletal System

The skeletal system is made of the **axial skeleton** and the **appendicular skeleton.** This chapter pertains to just the axial portion of the skeleton. The axial skeleton consists of the **skull,** the **thoracic cage,** and the **vertebral column.** Table 6-1 lists the details of the axial skeleton components.

Table 6-1 The Axial Skeleton	
Axial Skeleton	**Individual Components**
Skull	Cranium (8 bones)
	Face (14 bones)
	Auditory ossicles (6 bones)
	Hyoid (1 bone)
Thoracic cage	Sternum (1 bone)
	Ribs (24 bones)
Vertebral column	Vertebrae (24 bones)
	Sacrum (1 bone)
	Coccyx (1 bone)

The Skull

The skull is comprised of the **cranium,** the **facial bones,** the **ossicles,** and the **hyoid bone.** The following information discusses each component

Bones of the Cranium

Figure 6-1 identifies the cranial bones of the skull. There is one **frontal bone,** two **parietal bones,** and one **occipital bone.** There is one **sphenoid bone** (although it can be seen on both sides of the skull), and one **ethmoid bone.** There are two **temporal bones.**

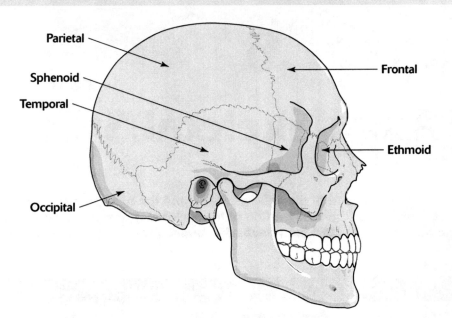

Figure 6-1: Lateral view of the cranial bones of the skull.

Example Problems

Use a directional term to answer Questions 1 through 3.

1. The lateral edge of the sphenoid bone is located _____ to the temporal bone.

    answer: anterior

2. The frontal bone is located _____ to the parietal bone.

    answer: anterior

3. The ethmoid bone is located on the _____ of the eye socket.

    answer: medial

4. True or false: The cheek bone (zygomatic) is part of the cranium.

    answer: false. It is part of the facial bone.

5. How many paired bones of the skull are there?

    answer: two (parietals and temporals)

Bones of the Face

Figures 6-2 through 6-4 identifies the facial bones. From an anterior view of the skull (see Figure 6-2), you can see the bones that make up the face. They are: two **nasal bones,** two **zygomatic bones,** one **mandible,** and two **maxillary bones.** (There is a suture that extends from the center of the nasal cavity to the front teeth that separates the two maxillary bones.) Inside the nasal cavity, you can see one **vomer bone** and two **inferior nasal conchae bones.** From a lateral view (Figure 6-3), you can see two **lacrimal bones.** From an inferior view (Figure 6-4), you can see two **palatine bones** (making up the posterior ⅓ of the roof of the mouth).

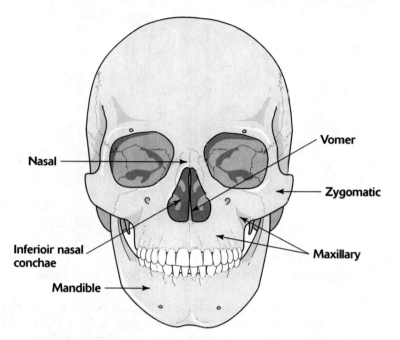

Figure 6-2: Anterior view of the facial bones of the skull.

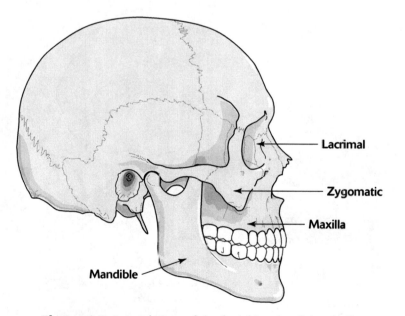

Figure 6-3: Lateral view of the facial bones of the skull.

The lacrimal bone is anterior to the **ethmoid bone.** The ethmoid bone is part of the cranium whereas the lacrimal bone (one on the other side, too) is part of the facial bones.

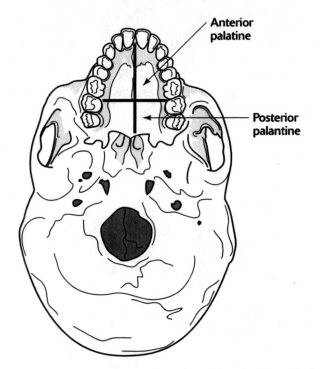

Figure 6-4: Inferior view of the facial bones of the skull.

The **anterior palatine** is not a separate bone. It is actually part of the maxillary bone. The **posterior palatine,** however, is a separate bone divided in the middle by a suture. Therefore, there are two posterior palatine bones. Figure 6-5 identifies some of the structures of the mandible.

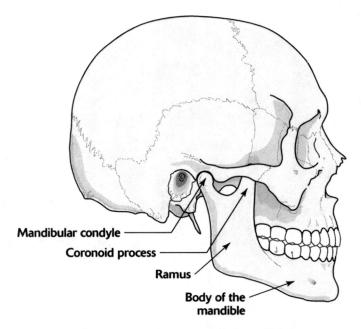

Figure 6-5: Features of the mandible.

Example Problems

1. The lower jaw is called the _____.

   answer: mandible

2. The upper jaw is called the _____.

   answer: maxilla

3. The cheek bones are called the _____.

   answer: zygomatic bones

4. The lacrimal bone is located on the medial side of the eye socket and is located _____ to the ethmoid bone.

   answer: anterior

5. How many paired bones make up the face? _____

   answer: 6 paired bones (maxilla, nasal, inferior nasal conchae, zygomatic, lacrimal, and palatine)

Internal Ear Bones

Within a bony ridge of the temporal bone (internal view of the skull) are **six ossicles.** There are three ossicles within each temporal bone. These are the six smallest bones of the body. These bones are connected to the **tympanic membrane** (ear drum) and to the hearing apparatus (**cochlea**) of the ear. Beginning with the attachment to the tympanic membrane is the **malleus, incus,** and then the **stapes.** In layman's terms, these are referred to as the **hammer, anvil,** and **stirrup.** The stapes actually does look like a tiny stirrup.

The Hyoid Bone

There is only one **hyoid bone.** It is located in the larynx region and is suspended by ligaments and is a place of attachment for muscles associated with the tongue and larynx.

Work Problems

1. Name the facial bone that makes up part of the nasal septum.

2. The anatomical name for the cheekbones is the _____.

3. The inferior nasal conchae are located (use a directional term) _____ to the vomer.

4. The anterior palatine structures are not separate bones. They are actually a part of the _____ bone.

5. When you put your hand on the back of your skull, your hand will be on the _____ bone.

6. There is one bone of the skull that does not connect to other bones. This bone is suspended by ligaments. Which bone is it?

7. When sound waves enter the ear canal, the waves will cause which ossicle to vibrate first?

8. The structure on the mandible that makes up part of the hinge joint of the jaw is called the _____.

9. The coronoid process of the mandible is located (use a directional term) _____ to the mandibular condyle.

10. Which ossicle is connected directly to the tympanic membrane?

Worked Solutions

1. **vomer**

2. **zygomatic**

3. **lateral**

4. **maxilla (maxillary)**

5. **occipital**

6. **hyoid.** This bone is suspended by ligaments.

7. **malleus**

8. **mandibular condyle**

9. **anterior**

10. **malleus**

Foramen of the Skull

There are numerous foramen in the skull. The **foramen** are holes in the bones that permit the passage of blood vessels and/or nerves to and from the brain. Figure 6-6 identifies some of the foramen of the skull located on the inferior side. Figure 6-7 identifies some of the foramen on the anterior side of the skull.

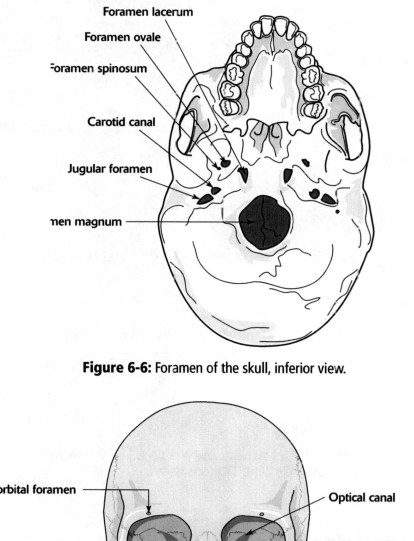

Foramen lacerum
Foramen ovale
Foramen spinosum
Carotid canal
Jugular foramen
men magnum

Figure 6-6: Foramen of the skull, inferior view.

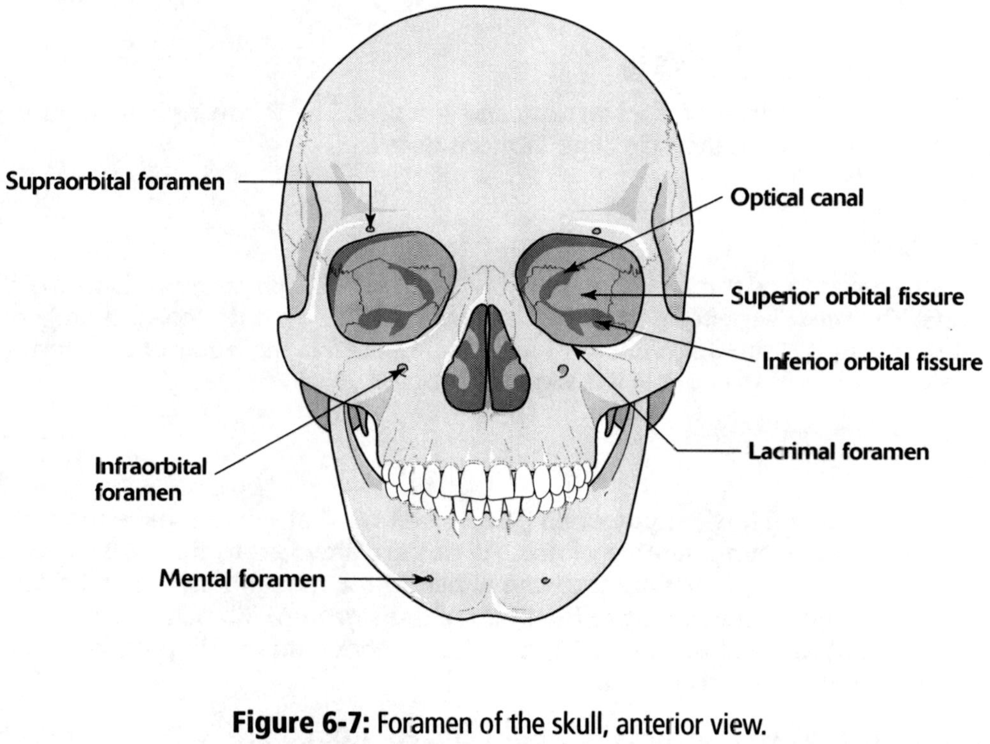

Supraorbital foramen
Optical canal
Superior orbital fissure
Inferior orbital fissure
Infraorbital foramen
Lacrimal foramen
Mental foramen

Figure 6-7: Foramen of the skull, anterior view.

Example Problems

1. The supraorbital foramen are located (use a directional term) _____ to the orbit of the eye.

   answer: superior

2. The infraorbital foramen are located on the _____ bone.

   answer: maxilla

3. The mental foramen are located on the _____ bone.

   answer: mandible

4. The jugular foramen are located mainly (use a directional term) _____ to the carotid canal.

   answer: posterior

5. The foramen spinosum are located mainly (use a directional term) _____ to the foramen ovale.

   answer: posterior

The Thoracic Cage

The **thoracic cage** is comprised of the **sternum** and the **ribs.** The following information discusses each component of the thoracic cage in more detail.

The Sternum

The sternum is referred to as the breast plate in layman's terms. The sternum is made of three individual parts. The most superior part is the **manubrium.** The **clavicle** (collar bone) connects to the manubrium and the shoulder. Inferior to the manubrium is the **body** of the sternum. The most inferior portion of the sternum is the **xiphoid** process.

The Ribs

There are 24 ribs (12 pairs). It is a myth when people believe that women have more ribs than men. Both sexes have the same number of ribs. All ribs are attached to the 12 thoracic vertebrae. The anterior portion of rib pair number one attach to the manubrium. Rib pairs number 2 through 7 have an anterior attachment to the body of the sternum. Rib pairs number 8 through 10 have an anterior attachment to the cartilage of the rib above them. Rib pairs number 11 and 12 do not have an anterior attachment at all.

The ribs are divided into two categories. Rib pair numbers 1 through 7 are called true ribs and rib pair numbers 8 through 12 are false ribs.

Anatomically, the ribs can also be divided in this manner: rib pair numbers 1 through 7 are **vertebrosternal ribs.** Rib pair numbers 8, 9, and 10 are called **vertebrochondral ribs.** These are the ribs that have an anterior attachment to the cartilage of the rib above them. Rib pair numbers 11 and 12 are called **floating ribs** because they do not have any anterior attachment.

The Vertebral Column

The vertebral column consists of 24 individual **vertebrae,** one **sacrum,** and one **coccyx.** The first seven vertebrae are called **cervical** vertebrae. These make up the bones of our neck. The vertebrae in the thoracic region are called the **thoracic** vertebrae. There are twelve of those. Each one has a pair of ribs attached to it. The last five vertebrae are the **lumbar** vertebrae.

When viewing the body from a lateral view, you will see that the vertebral column has four natural curves to it. The cervical region is an anterior curve. The thoracic region is a posterior curve. The lumbar region is an anterior curve. The sacrum and coccyx have a posterior curve. These curves are designed to support the body in an upright manner. Figure 6-8 shows how the curvatures help to support the body.

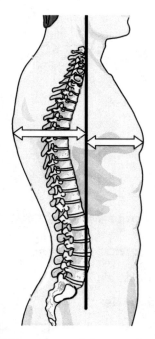

Figure 6-8: Curvatures of the vertebral column.

The line passing through the vertebrae represents the center of gravity. Because the line passes through the cervical vertebra, those vertebrae support the weight of the head. Notice that the thoracic vertebrae do not support the weight of the body. The thoracic vertebrae support the ribs. The rest of the line passes through the lumbar vertebrae. This means that the lumbar vertebrae support the weight of the body. This is one reason why the lumbar vertebrae are larger than the others.

One of the reasons why infants have a hard time standing up and walking is because their vertebral curvatures are not the same as they are in the adult. The infant's vertebral column is relatively straight, very little curvature. Therefore, there is more mass in front of the center of gravity compared to behind the center of gravity. The infant can topple over rather easy.

All vertebrae have these parts:

❏ Spine of the vertebrae

❏ Transverse processes

❏ Body

❏ Lamina

❏ Pedicle

❏ Vertebral foramen (the lamina and pedicle make up the boundaries of the vertebral foramen)

Figure 6-9 shows the parts of a typical thoracic vertebrae.

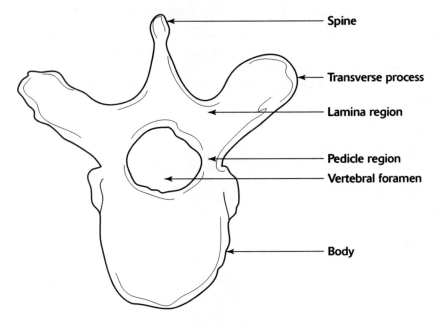

Figure 6-9: Vertebral parts.

The following are the features of the cervical vertebrae:

❏ Cervical number one is called the **atlas.**

❏ Cervical number two is called the **axis.**

❏ All cervical vertebrae have two additional foramen in addition to the vertebral foramen.

❏ Those extra foramen are called **transverse foramen.**

❏ The axis is the only vertebra that has a structure called a **dens.**

Here are the features of the thoracic vertebrae:

❏ The spine of the thoracic vertebrae angle inferiorly.

❏ There is a pair of ribs attached to each thoracic vertebra.

The lumbar vertebrae have two key features:

❏ The body of the lumbar vertebrae are larger than the other vertebrae.

❏ The spine of the lumbar vertebrae points straight posterior.

Example Problems

1. How many pairs of ribs do men have compared to women?

    answer: Both sexes have 12 pairs of ribs.

2. The ribs attach to which set of vertebrae?

    answer: thoracic vertebrae

3. How many pairs of ribs attach to the xiphoid process of the sternum?

    answer: Ribs do not attach to the xiphoid process. The ribs are attached to the manubrium and the body of the sternum.

4. What is the name of the vertebra that is attached to the skull?

    answer: atlas

5. True or false: The rib cage is part of the axial skeleton.

    answer: true

Work Problems

1. Cervical vertebrae can be identified from other vertebrae by the presence of extra holes called _____.

2. The dens (odontoid process) is the anterior projection on which cervical vertebra?

3. Ribs that do not have a connection to the sternum are called _____.

4. How many pairs of true ribs are there and of those true ribs, how many are considered to be vertebrosternal?

5. How many pairs of false ribs are there and of those false ribs, how many are considered to be vertebrochondral?

6. The spinal cord passes through a large hole in the vertebrae called the _____.

7. The vertebral column has how many cervical vertebrae?

8. Blood vessels and nerves pass through _____ of the skull to get to their destination.

9. The largest foramen of the skull is the foramen _____.

10. The foramen spinosum is mostly (use a directional term) _____ to the foramen ovale.

Worked Solutions

1. **transverse foramen**

2. **vertebra number 2; the axis vertebra**

3. **false ribs.** This would be a combination of the vertebrochondral ribs and floating ribs.

4. **There are 7 pairs of true ribs and all 7 pairs are vertebrosternal.** These have a more direct attachment to the sternum.

5. **There are 5 pairs of false ribs and 3 pairs (Numbers 8–10) are vertebrochondral.** These attach to the cartilage of the rib superior to it.

6. **vertebral foramen**

7. **7**

8. **foramen**

9. **magnum**

10. **posterior**

Chapter Problems and Solutions

Problems

1. In an adult, there are _____ natural curves in the vertebral column as seen from a lateral view.

2. What is the name of the bony structure that makes up part of the nasal septum?

3. How many vertebrae are there (not counting the sacrum or coccyx)?

4. The clavicle articulates with which part of the sternum?

5. The jugular foramen is located mainly (use a directional term) _____ to the carotid canal (foramen).

6. True or false: There are 24 vertebrae and 24 ribs; there is one rib per vertebrae.

7. The coccyx (tail bone) is attached to the _____.

8. There are bony structures in the nasal cavity that are designed to cause air to swirl around before entering the trachea. This swirling action causes the air to warm up. What are those structures called?

9. When a person gets whiplash, it is because they damaged which set of vertebrae?

10. The area located between the transverse processes and the spinous process is called the _____.

Answers and Solutions

1. **4**

2. **vomer**

3. **7 cervical, 12 thoracic, and 5 lumbar = 24**

4. **It articulates with the manubrium of the sternum.**

5. **posterior**

6. **false.** There are two ribs attached to each thoracic vertebrae (for a total of 24 ribs).

7. **It attaches to the sacrum.**

8. **nasal conchae**

9. **cervical vertebrae**

10. **lamina**

Supplemental Chapter Problems

Problems

1. Spina bifida is a condition where the walls of the vertebral foramen do not form properly. The developing child's spinal cord material will protrude through the opening that is formed by this malformation. What two structures make up the wall of the vertebral foramen?

2. Calvaria is a term that refers to the skull cap. True or false: the zygomatic bones are part of the calvaria.

3. The foramen magnum is located in what bone of the skull?

4. The hinge of the lower jaw is called the _____.

5. The roof of the mouth is made of the maxillary bone and also the _____ bone.

6. The sacrum attaches to which vertebrae?

7. True or false: The thoracic vertebrae support the weight of the body.

8. How many pairs of false ribs are there?

9. How many pairs of vertebrosternal ribs are there?

10. Which term is singular: vertebra or vertebrae?

Answers

1. the lamina and pedicle

2. false. The zygomatic bone is part of the facial bones.

3. occipital bone

4. mandibular condyle

5. posterior palatine bone

6. lumbar vertebrae

7. false. The lumbar vertebrae support the weight of the body. The thoracic vertebrae support the ribs.

8. There are 5 pairs of false ribs.

9. There are 7 pairs of vertebrosternal ribs.

10. vertebra

Chapter 7

The Appendicular Skeletal System

The skeletal system is made of the **axial skeleton** and the **appendicular skeleton.** This chapter pertains to just the appendicular portion of the skeleton, which consists of the **pectoral girdle** (shoulder), the **upper limbs,** the **pelvic girdle** (hip), and the **lower limbs.**

The Pectoral Girdle and Upper Limbs

Table 7-1 lists the details of the appendicular skeleton components of the upper body.

Table 7-1 The Appendicular Skeleton—Upper Body		
Appendicular Skeleton	*Upper Limbs*	*Individual Components*
Pectoral girdle	Scapula and clavicle	Spinous process
		Acromion
		Coracoid process
		Glenoid cavity
Upper limbs	Humerus	Head
		Greater tubercle
		Medial and lateral epicondyle
		Capitulum
		Trochlea
		Coronoid fossa
		Olecranon fossa
	Radius	Head
		Dorsal radial tuberosity
		Styloid process

(continued)

Table 7-1 The Appendicular Skeleton—Upper Body *(continued)*		
Appendicular Skeleton	**Upper Limbs**	**Individual Components**
	Ulna	Olecranon process
		Trochlear notch
		Styloid process
	Wrist (carpals)	Capitate
		Hamate
		Pisiform
		Triquetrum
		Lunate
		Scaphoid
		Trapezium
		Trapezoid
	Hand and fingers	Metacarpals
		Phalanges

❑ **Clavicle:** The clavicle connects to the **manubrium of the sternum** and the **acromion of the scapula.** Many times, you can feel a bump on the top, lateral edge of your shoulder. This bump is the union of the clavicle to the acromion of the scapula.

❑ **Scapula:** You can feel the **spinous process** of the scapula. It is the posterior ridge that angles from the acromion to the medial edge of the scapula. The **glenoid cavity** is at the lateral edge of the scapula and is the socket portion of the ball-and-socket joint of the shoulder. The **acromion** of the scapula connects to the clavicle.

❑ **Humerus:** The head of the humerus fits into the glenoid cavity of the scapula. Lateral to the head is the **greater tubercle.** You can feel the greater tubercle on your body by placing your finger on the lateral, superior edge of the humerus. At the inferior (distal) end of the humerus are two **condyles.** These condyles have special names. The lateral condyle is the **capitulum** and the medial condyle is the **trochlea.** Lateral to the capitulum is a rather large bump called the lateral epicondyle. Medial to the trochlea is the **medial epicondyle.** You can feel these epicondyles on your own body. There is a groove between the medial epicondyle and the trochlea. When people hit this area, they say they've hit their "funny bone." There is a nerve that passes through that area. This nerve is activated when it is hit. On the anterior side of the humerus, at the distal end, there is a depression called the **coronoid fossa.** On the opposite side is a larger depression called the olecranon fossa.

❑ **Radius:** There are two bones comprising the lower arm. The radius is the lateral bone and the ulna is the medial bone. When the hand is in the supinate position, the radius and ulna are parallel to each other. When the hand is pronated, the radius crosses over the ulna. The head of the radius pivots on the capitulum. At the inferior (distal) end of the radius, there are bumps on the posterior side. These bumps are the **dorsal radial tuberosity.** The anterior side, at the distal end, is smooth, not bumpy like the posterior side.

❏ **Ulna:** The ulna has a large bulge on the posterior side called the **olecranon process.** This is the elbow. Anterior to the olecranon process is a huge notch called the **trochlear notch.** The trochlear notch pivots on the trochlea on the ulna.

❏ **Wrist:** The wrist is made of the eight carpal bones. The **scaphoid** hinges with the radius and the lunate hinges with both the radius and the ulna. The **capitate** is a rather large bone that is "in line" with metacarpal III and the middle finger. Medial to the capitate is the **hamate.** On the anterior side (palm side) the hamate has a hook. Medial to the hamate are two bones; the **pisiform** and **triquetrum.** The pisiform is a small, round bone sitting on the triquetrum. Lateral to the pisiform (nearest the ulna) is the **lunate.** Lateral to the lunate (nearest the radius) is the **scaphoid.** Nearest metacarpal I and in line with the thumb is the **trapezium.** Between the trapezium and the capitate is the **trapezoid.**

❏ **Metacarpals:** The "back of the hand" consists of the metacarpals. The metacarpals are numbered with Roman numerals beginning the most lateral metacarpal. Metacarpal I is in line with the pollex. The most inferior (distal) portion of the metacarpals makes up the "knuckles" of the hands.

❏ **Digits:** The fingers and thumb are made of **phalanges.** The thumb has two phalanges; **proximal phalange** and **distal phalange.** The other digits have three phalanges: **proximal phalange, middle phalange,** and **distal phalange.**

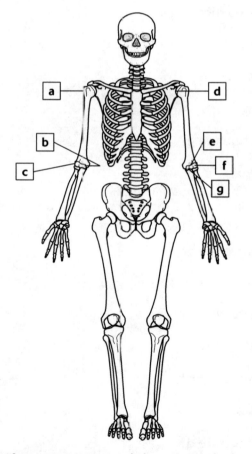

Figure 7-1: Anterior skeleton, upper body.

Example Problems

Use Figure 7-1 and the terms in Table 7-1 to answer the following questions in reference to the anterior skeleton.

1. What is the name for structure a?

   answer: greater tubercle

2. What is the name for structure b?

   answer: medial epicondyle

3. What is the name for structure c?

   answer: lateral epicondyle

4. What is the name for structure d?

   answer: head of the humerus

5. What is the name for structure e?

   answer: coronoid fossa

6. What is the name for structure f?

   answer: trochlea

7. What is the name for structure g?

   answer: capitulum

Use Figure 7-2 and the terms in Table 7-1 to answer the following questions in reference to the posterior skeleton.

8. What is the name for structure a?

   answer: spinous process of the scapula

9. What is the name for structure b?

   answer: acromion of the scapula

10. What is the name for structure c?

    answer: olecranon fossa

11. What is the name for structure d?

    answer: dorsal radial tuberosity

12. What is the name for structure e?

    answer: olecranon process

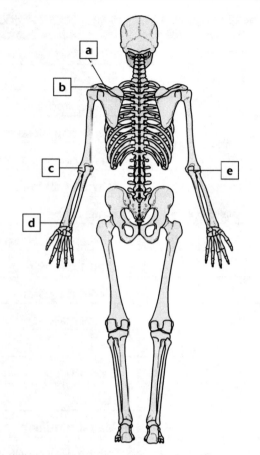

Figure 7-2: Posterior skeleton, upper body.

The Pelvic Girdle and Lower Limbs

Table 7-2 lists the details of the appendicular skeleton components of the lower body, which are also shown in Figure 7-3.

Table 7-2 The Appendicular Skeleton—Lower Body		
Appendicular Skeleton	*Lower Limbs*	*Individual Components*
Pelvic girdle	Hip	Ilium
		Pubis
		Ischium
		Acetabulum
		Greater sciatic notch
		Obturator foramen

(continued)

Table 7-2 The Appendicular Skeleton—Lower Body *(continued)*		
Appendicular Skeleton	*Lower Limbs*	*Individual Components*
Lower limbs	Femur	Head
		Greater trochanter
		Medial and lateral epicondyle
		Medial and lateral condyle
		Intercondylar fossa
	Patella	(Knee cap)
	Tibia	Intercondylar tubercles
		Tibial tuberosity
		Medial malleolus
	Fibula	Lateral malleolus
	Ankle (tarsals)	Talus
		Calcaneus
		Navicular
		Medial cuneiform
		Intermediate cuneiform
		Lateral cuneiform
		Cuboid
	Foot and Toes	Metatarsals
		Phalanges

❑ **Hip:** A complete hip is made of two large bones called **os coxae.** Each os coxae consists of the **ilium, pubis,** and **ischium.** The ilium consists of the **iliac crest,** which is the superior edge of the hip. The pubis is the area where the two os coxae join together anteriorly. Between the pubis bones of the two os coxae is a pad of cartilage called the **pubic symphysis.** The portion of the hip that we sit on is the ischium. The lateral portion of each os coxae has a deep fossa called the **acetabulum.** The acetabulum is the socket portion of the ball-and-socket joint of the hip. On the posterior edge of each os coxae is a huge notch called the **greater sciatic notch.** The sciatic nerve passes through this notch. Inferior to the acetabulum is a huge hole called the **obturator foramen.**

❑ **Femur:** The head of the femur fits into the acetabulum. Lateral to the head is a huge bulge called the **greater trochanter.** You can feel the greater trochanter on your body by placing your hands in the area of your pant's pockets. At the inferior (distal) end of the femur are two condyles. The **lateral condyle** is on the same side of the femur as the greater trochanter. The **medial condyle** is on the same side of the femur as the head. The two bulges on either side of the condyles are the **epicondyles.** You can feel the epicondyles as bulges on either side of the patella. Between the two condyles is a large depression called the **intercondylar fossa.** A portion of the anterior and posterior cruciate ligaments (ACL and PCL) attach in the intercondylar fossa region.

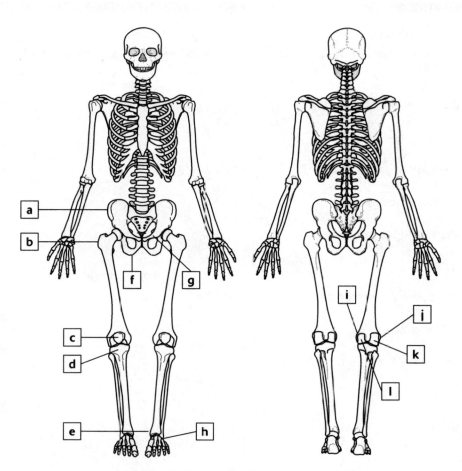

Figure 7-3: Anterior skeleton, lower body.

❑ **Tibia:** The lower leg consists of two parallel bones. The lateral bone is the fibula and the medial bone is the tibia. The tibia hinges with the femur making a hinge joint. The superior top of the tibia consists of two bumps called the **intercondylar tubercles.** The ACL and PCL also attach to the intercondylar tubercles. On the anterior side (near the superior end) of the tibia is a roughened bulge called the **tibial tuberosity.** At the distal end, on the medial side is a bulge. This bulge appears to be on the medial side of your ankle. It really isn't a part of the ankle; it's a part of the tibia. This bulge is called the medial malleolus.

❑ **Fibula:** At the distal end of the fibula, on the lateral side, is the lateral malleolus.

❑ **Ankle:** There are seven tarsal bones comprising the ankle. The tibia pivots on the talus. The large heel bone is the **calcaneus.** Anterior to the talus is the **navicular.** Anterior to the navicular are three **cuneiform bones.** The cuneiform bones are identified by position. The cuneiform bone on the medial side is the **medial cuneiform.** Lateral to it is the **intermediate cuneiform.** Lateral to it is the **lateral cuneiform.** Lateral to it is the cuboid bone.

❑ **Metatarsals:** The arch of the foot consists of the metatarsal bones. The metatarsals are identified with Roman numerals. Metatarsal number I is the most medial bone. Metatarsal I is in line with the hallux.

❑ **Digits:** The toes and hallux are made of **phalanges.** The hallux has two phalanges: **proximal phalange** and **distal phalange.** The other digits have three phalanges: **proximal phalange, middle phalange,** and **distal phalange.**

Example Problems

Use Figure 7-3 and the terms in Table 7-2 to answer the following questions in reference to the anterior skeleton.

1. What is the name for structure a?

    answer: ilium

2. What is the name for structure b?

    answer: greater trochanter

3. What is the name for structure c?

    answer: patella

4. What is the name for structure d?

    answer: tibial tuberosity

5. What is the name for structure e?

    answer: medial malleolus

6. What is the name for structure f ?

    answer: ischium

7. What is the name for structure g?

    answer: pubis

8. What is the name for structure h?

    answer: lateral malleolus

9. What is the name for structure i?

    answer: medial epicondyle

10. What is the name for structure j?

    answer: lateral epicondyle

11. What is the name for structure k?

    answer: lateral condyle

12. What is the name for structure l?

    answer: medial condyle

Work Problems

1. The bones that make up the wrist are collectively called _____.

2. The bones that make up the ankle are collectively called _____.

3. What is the name of the bones that are located between the wrist bones and the knuckles?

4. What is the name of the bones that are located between the ankle bones and the toes?

5. Describe the location of the fibula.

6. Describe the location of the radius.

7. The head of the femur would be considered a (lateral or medial) structure of the femur.

8. The tibial tuberosity is located on the _____ side of the tibia.

9. The head of the femur and the _____ make up the ball-and-socket joint of the leg.

10. Identify the bone that consists of the lateral malleolus.

Worked Solutions

1. **carpals**

2. **tarsals**

3. **metacarpals**

4. **metatarsals**

5. **The fibula is the lateral bone of the lower leg.**

6. **The radius is the lateral bone of the antebrachium.**

7. **medial**

8. **anterior**

9. **acetabulum**

10. **fibula**

Chapter Problems and Solutions

Problems

1. Which end of the clavicle attaches to the sternum, the medial end or the lateral end?

2. On your own body, can you feel the medial epicondyle or the capitulum of the humerus?

3. Is the distal, anterior portion of the radius smooth or rough?

4. How many bones make up the pectoral girdle and upper limbs for the entire body?

5. The acetabulum is a medial or a lateral structure of the os coxae?

6. A large fossa on the posterior side of the humerus is the _____.

7. Each os coxae consists of three fused bones. They are the _____, the _____, and the _____.

8. How many bones make up the pelvic girdle and lower limbs for the entire body?

9. The Achilles' heel is a layman's term. In anatomy, this tendon is actually the calcaneal tendon. It is called the calcaneal tendon because it attaches to which of the tarsal bones?

10. The sacrum provides the attachment for which of the os coxae structures?

Answers and Solutions

1. **The medial end of the clavicle attaches to the manubrium of the sternum.**

2. **You can feel the medial epicondyle of the humerus.** The capitulum is covered by muscle tissue.

3. **The distal, anterior portion of the radius is smooth.** The distal, posterior portion has bumps on it called the dorsal radial tuberosity.

4. **There are 2 clavicles.** There are 2 scapulae. There are 2 humerus bones, 2 radial bones, and 2 ulnar bones. There are 8 carpals per wrist and 5 metacarpals per hand. There are 14 phalanges per hand as well. The total number is 64 individual bones.

5. **The acetabulum is a lateral structure.** It is lateral so the head of the femur, which is a medial structure of the femur, can fit into the socket.

6. **The large fossa is called the olecranon fossa.**

7. **The three fused bones are: the ilium, the ischium, and the pubis.**

8. **There are 2 os coxae, 2 femurs, 2 patellar bones, 2 tibial bones, and 2 fibular bones.** There are 7 tarsals per ankle and 5 metatarsals per foot. The toes consist of 14 bones per foot. The total number is 62 individual bones.

9. **This tendon attaches to the calcaneus, hence the anatomical name for the tendon.**

10. **The ilium of the os coxae connects to the sacrum.** The pubis bones of the os coxae are attached to each other via the pubic symphysis.

Supplemental Chapter Problems

Problems

1. The spinous process of the scapula is an anterior or a posterior structure?

2. The glenoid cavity of the scapula is a lateral or a medial structure?

3. What is the name of the bony structure that is actually the elbow?

4. The capitate is medial or lateral to the trapezoid?

5. Metacarpal number _____ is associated with the little finger side?

6. The crest of the hip is part of which bony structure of the os coxae?

7. The head of the femur is a medial or a lateral structure of the femur?

8. The medial malleolus is on which bone of the lower leg?

9. The lateral cuneiform is medial to which tarsal bone?

10. The greater trochanter is a medial or lateral structure of the femur?

Answers

1. posterior

2. lateral

3. olecranon process

4. medial

5. V (Roman numerals are used)

6. ilium

7. medial

8. tibia

9. cuboid

10. lateral

Chapter 8
The Muscular System

E very action the body takes utilizes a muscular activity. Some of the muscles of the body are under **voluntary control** (skeletal muscles), and by using these muscles, you are able to respond to situations that, therefore, lead to survival. Some muscles are under **involuntary control.** These muscles respond to situations automatically, which also leads to survival. The smooth muscles typically are associated with the internal organs of the body, whereas the skeletal muscles are associated with the muscles attached to the skeleton (such as the arms and legs). There is one tissue that is made of **cardiac muscle,** and that is the heart. The muscle of the heart contracts involuntary but has a rhythmic pattern. This chapter discusses the skeletal muscles.

Select Muscles of the Face

Table 8-1 lists select muscles of the face with a brief description of their location. Figure 8-1 shows the select muscles of the face.

Table 8-1 Select Muscles of the Face	
Muscle	*Description of Location*
Frontalis	Located on the forehead
Occipitalis	Located on the occipital bone
Temporalis	Located on the temporal bone
Orbicularis oculi	Encircles the eye
Orbicularis oris	Encircles the mouth
Zygomaticus	Extends from the corner of the mouth to the zygomatic bone
Risorius	Extends from the corner of the mouth straight posterior
Masseter	Located on the ramus of the mandible

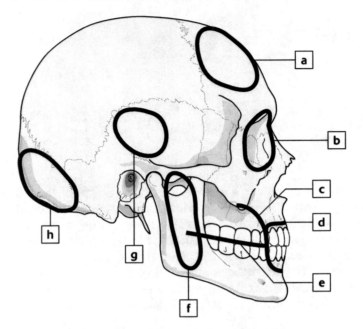

Figure 8-1: Select muscles of the face.

Example Problems

Use the description of the muscle location from Table 8-1 to identify the muscles on Figure 8-1.

1. Muscle a is the _____.

    answer: frontalis

2. Muscle b is the _____.

    answer: orbicularis oculi

3. Muscle c is the _____.

    answer: zygomaticus

4. Muscle d is the _____.

    answer: orbicularis oris

5. Muscle e is the _____.

    answer: risorius

6. Muscle f is the _____.

    answer: masseter

7. Muscle g is the _____.

    answer: temporalis

8. Muscle h is the _____.

    answer: occipitalis

Select Muscles of the Arm

Table 8-2 lists select muscles of the arm with a brief description of their location. Figure 8-2 shows the select muscles of the arm.

Table 8-2 Select Muscles of the Arm	
Muscle	**Description of Location**
Biceps brachii	Located on the anterior upper arm.
Triceps brachii	Located on the posterior upper arm.
Palmaris longus	Located on the lower arm and extends to the center of the palm.
Extensor digitorum	Located on the lower arm and extends to the center of the back of the hand.
Pronator teres	Located in the antecubital region. It pronates the hand by crossing the radius over the ulna.
Supinator	Located in the cubital region. It supinates the hand by making the lower arm bones parallel to each other.
Brachioradialis	Located lateral to the elbow on the lower arm.
Deltoid	Located on the shoulder and attaches to the upper arm.

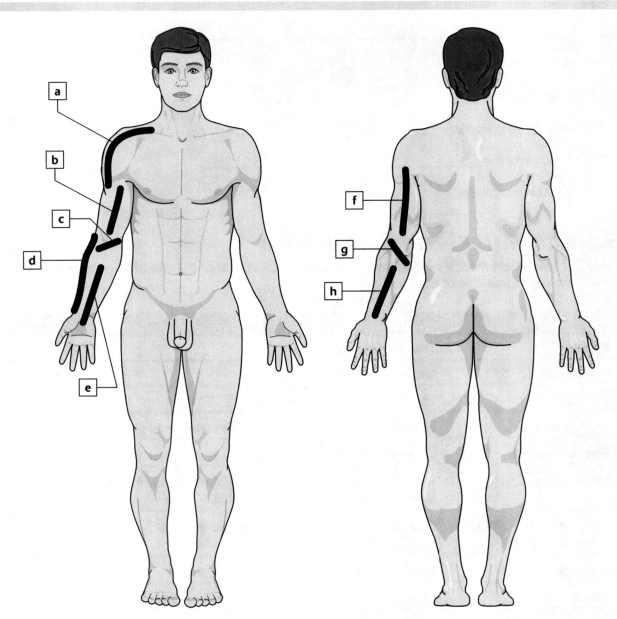

Figure 8-2: Select muscles of the arm.

Example Problems

Use the description of the muscle location from Table 8-2 to identify the muscles on Figure 8-2.

1. Muscle a is the _____.

    answer: deltoid

2. Muscle b is the _____.

    answer: biceps brachii

3. Muscle c is the _____.

    answer: pronator teres

4. Muscle d is the _____.

    answer: brachioradialis

5. Muscle e is the _____.

    answer: palmaris longus

6. Muscle f is the _____.

    answer: triceps brachii

7. Muscle g is the _____.

    answer: supinator

8. Muscle h is the _____.

    answer: extensor digitorum

Select Muscles of the Leg

Table 8-3 lists select muscles of the leg with a brief description of their location. Figure 8-3 shows the select muscles of the leg.

Table 8-3 Select Muscles of the Leg	
Muscle	*Description of Location*
Sartorius	Angles across the anterior side of the thigh.
Rectus femoris	Located in the center of the anterior thigh.
Vastus medialis	Located medial to the rectus femoris.
Vastus lateralis	Located lateral to the rectus femoris.
Vastus intermedius	Located posterior to the rectus femoris (under the rectus femoris).
Biceps femoris	Located on the posterior thigh. It is the lateral of two parallel muscles on the posterior thigh.
Semitendinosus	Located on the posterior thigh. It is the medial of the two parallel muscles on the posterior thigh.
Gracilis	The most medial muscle of the thigh.
Tibialis anterior	Located on the anterior lower leg.
Gastrocnemius	Located on the posterior lower leg.

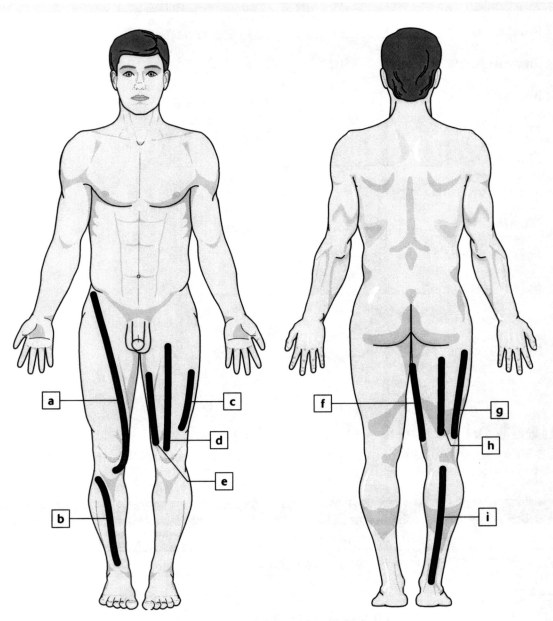

Figure 8-3: Select muscles of the leg.

Example Problems

Use the description of the muscle location from Table 8-3 to identify the muscles on Figure 8-3.

1. Muscle a is the _____.

    answer: sartorius

2. Muscle b is the _____.

    answer: tibialis anterior

3. Muscle c is the _____.

    answer: vastus lateralis

4. Muscle d is the _____.

    answer: rectus femoris

5. Muscle e is the _____.

    answer: vastus medialis

6. Muscle f is the _____.

    answer: gracilis

7. Muscle g is the _____.

    answer: biceps femoris

8. Muscle h is the _____.

    answer: semitendinosus

9. Muscle i is the _____.

    answer: gastrocnemius

10. In Figure 8-3, identify the muscle letter to which the vastus intermedius is posterior.

    answer: d. It is located posterior to the rectus femoris muscle.

Select Muscles of the Torso

Table 8-4 lists select muscles of the torso with a brief description of their location. Figure 8-4 shows the select muscles of the torso.

Table 8-4 Select Muscles of the Torso	
Muscle	*Description of Location*
Pectoralis major	Located on the chest.
Trapezius	Located on the upper back.
Rectus abdominis	Vertical muscles on the abdomen.
Latissimus dorsi	Located on the lower back.
External oblique	Located lateral to the rectus abdominis.
Gluteus maximus	Located in the gluteal region.
Sternocleidomastoid	Extends from the manubrium of the sternum to the mastoid process of the skull.

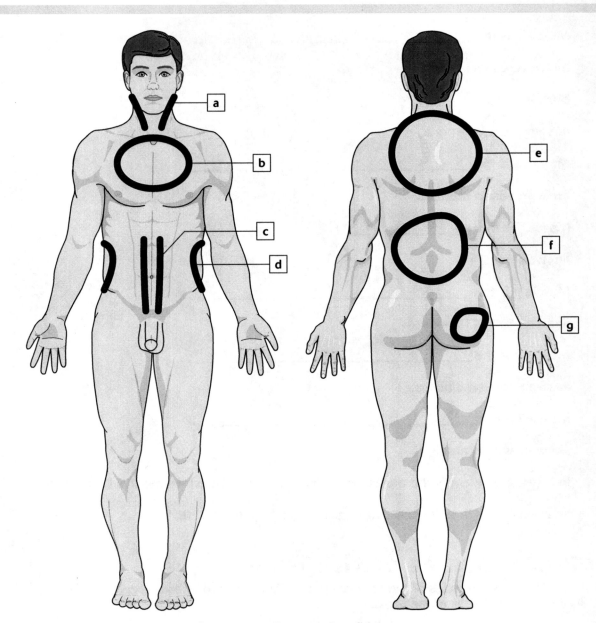

Figure 8-4: Select muscles of the torso.

Example Problems

Use the description of the muscle location from Table 8-4 to identify the muscles on Figure 8-4.

1. Muscle a is the _____.

    answer: sternocleidomastoid

2. Muscle b is the _____.

    answer: pectoralis major

3. Muscle c is the _____.

   answer: rectus abdominis

4. Muscle d is the _____.

   answer: external oblique

5. Muscle e is the _____.

   answer: trapezius

6. Muscle f is the _____.

   answer: latissimus dorsi

7. Muscle g is the _____.

   answer: gluteus maximus

Work Problems

Identify the muscle being described. Use the muscles that are identified in Tables 8-1 through 8-4.

1. This muscle is located on the frontal bone.

2. This muscle is located in the center of the anterior antebrachium.

3. This muscle is located on the crus (crural region).

4. This muscle is located on the sura (sural region).

5. This muscle is located lateral to the gluteal cleft.

6. This is a rather large muscle located on the lateral side of the elbow and radius.

7. This muscle is located in the center of the anterior thigh.

8. When this muscle contracts, it will cause the arm to move laterally.

9. This muscle is located inferior to the trapezius.

10. This muscle is located lateral to the rectus abdominis.

Worked Solutions

1. **frontalis**

2. **palmaris longus**

3. **tibialis anterior**

4. **gastrocnemius**

5. **gluteus maximus**

6. **brachioradialis**

7. **rectus femoris**

8. **deltoid**

9. **latissimus dorsi**

10. **external oblique**

Muscle Structure

In order to understand how muscles contract, one must understand the structure of muscles. The following information starts with the macroscopic level of the muscle and ends with the microscopic level.

Consider the biceps brachii as an example. The entire, main body of the muscle is made of several **muscle fasciculi.** Each fasciculus is made of **muscle fibers** (a muscle fiber is the same as a muscle cell). Each muscle fiber is made of myofibrils. The **myofibrils** are protein molecules (actin and myosin).

Upon examination at the protein level, it is found that the myofibrils are arranged in a specific manner. Figure 8-5 shows this arrangement.

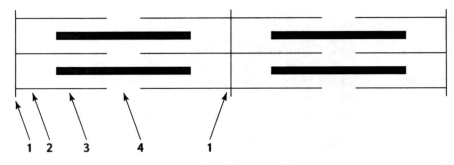

Figure 8-5: Muscle sarcomeres.

The thick, dark lines represent **myosin.** The thin lines are **actin.** Notice that actin and myosin overlap each other is certain areas. Area number 1 is called the Z line. Area number 2 is called the I band. The I band consists of actin, space, actin, space, actin. Area number 3 is called the A band. The A band consists of overlapping actin and myosin. It consists of actin, myosin, actin, myosin, actin. Area number 4 is called the H band. It consists of: space, myosin, space, myosin, and so on. The span between one Z line and the next Z line is called a sarcomere. Figure 8-6 shows two sarcomeres. Each sarcomere is a functional unit of muscle.

Researchers noticed that when a muscle was stimulated to contract, the "empty space" of the I band got shorter. They began to study to determine what caused the I band to get shorter. It was found that the myosin myofilament also consisted of structures called **cross-bridges.** The cross-bridges extended (upon stimulation) and attached to actin. Once they attached to actin, they pivoted, thereby causing actin myofilament to move toward each other. The myosin myofilaments are stationary; only the actin moves. Because of the movement of the actin, this process is called **the sliding filament theory.** Figure 8-6 shows the cross-bridges associated with the myosin filaments. Those cross-bridges stretch and attach to the actin myofilament. Figure 8-7 shows the cross-bridges already attached to actin and in the pivoted position.

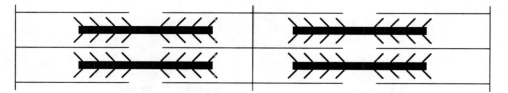

Figure 8-6: Muscle sarcomeres showing cross-bridges.

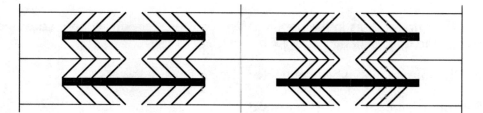

Figure 8-7: Muscle sarcomeres showing cross-bridge attachment and pivot.

Notice that the I band got shorter, and the H band got shorter. The area of the overlap of actin and myosin increased. The Z lines are closer together. The shortening of muscle sarcomeres is muscle contraction. When a muscle relaxes, the cross-bridges turn loose, and the actin myofilament slide back to their original position.

Located on the actin filament are **binding sites** for the cross-bridges to attach. These binding sites are originally blocked by two other protein filaments. One is called **troponin** and the other is called **tropomyosin.** These two are typically referred to as the **troponin/tropomyosin complex.** This complex must move out of the way to allow cross-bridges to bind to actin. It has been found that **calcium ions** cause the troponin/tropomyosin molecules to move out of the way, thus exposing the binding sites so the cross-bridges can attach.

There is a cell organelle in the muscle cell called the **sarcoplasmic reticulum.** This organelle stores calcium ions. A nerve impulse ultimately causes the sarcoplasmic reticulum to release the calcium ions. The calcium ions then cause the troponin/tropomyosin complex to move out of the way to allow cross-bridges to bind to actin. Once the cross-bridges have attached to actin, ATP is used to cause the cross-bridges to pivot, thus muscle contraction.

Example Problems

1. How many sarcomeres are depicted in Figure 8-6?

   answer: two

2. When the cross-bridges pivot, they will cause the _____ myofilament to move.

   answer: actin. Myosin is stationary.

3. What ion has been found to be directly involved in muscle contraction?

   answer: calcium ions (Ca^{2+})

4. Upon contraction, the H band and the _____ band gets shorter.

   answer: I band

5. Cross-bridges are a part of which myofilament?

   answer: myosin

Chapter Problems and Solutions

Problems

1. You can feel two tendons in the popliteal region. These tendons connect posterior thigh muscles to the tibia. What are those two muscles?

2. Identify the muscle that, when contracted, would flex the lower arm.

3. The area of the sarcomere that consists of just myosin filaments is the _____ band.

4. The area of the sarcomere that consists of overlapping actin and myosin filaments is the _____ band.

5. When a muscle is not contracting, the binding sites on actin are blocked by what molecular complex?

6. The sarcoplasmic reticulum stores _____ ions.

7. Identify the three types of muscle tissue in the body. _____, _____, and _____.

8. Identify the muscle that can be found in the sural region. _____

9. In order to move the arm laterally, the muscle on the shoulder has to contract. What is the name of that muscle?

10. Which muscle do trumpet players utilize?

Answers and Solutions

1. **biceps femoris and semitendinosus**

2. **biceps brachii**

3. **the H band**

4. **the A band**

5. **troponin/tropomyosin complex**

6. **calcium**

7. **skeletal, smooth, and cardiac**

8. **gastrocnemius**

9. **deltoid**

10. **orbicularis oris**

Supplemental Chapter Problems

Problems

1. Based on the location of the muscles, which muscle, from Table 8-1, allows you to raise one corner of your lip?

2. Which muscle, located on the mandible (refer to Table 8-1) allows you to close your mouth?

3. Based on the location of the muscles, which muscle in Table 8-2 allows you to flex your wrist?

4. Based on the location of the muscles, which muscle in Table 8-2 allows you to place your palms in the anatomical position?

5. Based on the location of the muscles, which muscle in Table 8-3 pulls on the calcaneus thus allowing you to stand on your tippy-toes?

6. Based on the location of the muscles, which two muscles in Table 8-3 allows you to flex your lower leg?

7. Based on the location of the muscles, which muscle in Table 8-4 allows you move your entire leg posteriorly?

8. Based on the location of the muscles, which muscle in Table 8-4 allows you tilt your head backward?

9. The troponin/tropomyosin complex molecules are mainly associated with which myofilament?

10. What ion is necessary for muscle contraction to occur?

Answers

1. zygomaticus

2. masseter

3. palmaris longus

4. supinator

5. gastrocnemius

6. semitendinosus and biceps femoris

7. gluteus maximus

8. sternocleidomastoid

9. actin

10. calcium ion

Chapter 9
General Nervous System

The nervous system is very complex and is, therefore, subdivided into two major groups. One is the central nervous system (CNS), and the other is the peripheral nervous system (PNS). In order to understand the nature of the CNS, you must first gain an understanding of the basic structure of the main functional unit of the nerves. This chapter discusses the structure of the neurons.

Neurons

The cells that comprise the nervous system are of two major types; one is called a **neuron** and the other is the **neuroglia.** Table 9-1 lists the parts of a neuron with a brief description of each part.

Table 9-1 Neuron Information	
Neuron Structure	**Brief Description**
Soma	This is the main cell body of the neuron.
	It consists of a nucleus and cell organelles.
	Does not have any centrioles and therefore lacks the ability to reproduce or repair.
Axon	This is the long, single extension of the soma.
	The end of the axon will innervate with a muscle or gland for example.
	Axons in the peripheral nervous system can repair slowly.
Dendrite	These are short and variable extensions of the soma on the opposite side of the axon.
Presynaptic vesicle	This is a vesicle located at the end of the axon.
	It consists of neurotransmitters.
	One main neurotransmitter is acetylcholine.

The membrane of the dendrites and soma are sensitive to changes in the environment. Once the membranes have been stimulated, an impulse may occur. The impulse will travel down the axon to its destination.

Figure 9-1 shows a simplified view of a neuron and an impulse traveling through it.

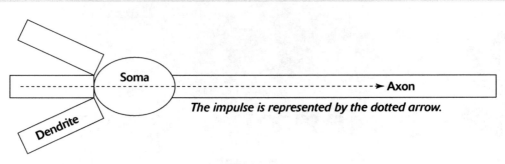

The impulse is represented by the dotted arrow.

Figure 9-1: Simplified neuron.

The Impulse

The outer surface of the neuron membrane consists of numerous positive ions. These are mainly sodium ions (Na^+) and calcium ions (Ca^{2+}). In many cases, there may be several neurons in sequence. Between each neuron, there is a small gap called the synapse. A neuron that consists of positive ions on the outside of the membrane is said to be a **polarized nerve.** Figure 9-2 shows a polarized neuron and the synapse.

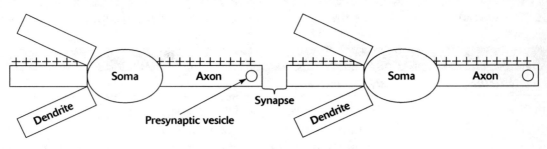

Figure 9-2: Polarized neuron.

When the membrane of the dendrite is activated by a stimulus, Na^{1+} will enter. These ions will diffuse along the membrane of the dendrite and soma (in the ICF region). When the ions reach the junction of the soma and axon (an area called the axon hillock), the membrane becomes very permeable to Na^{1+}. There is a tremendous influx of Na^{1+} in the axon. The ions in the ECF of the axon will enter the axon in a domino fashion. Therefore, when one positive ion enters (the process is called **depolarization**) an adjacent positive will then enter. This will then cause the next positive ion to enter and so forth. As each ion enters the neuron, a wave-like activity occurs. This wave-like activity is the impulse. This traveling wave of activity is known as the **action potential.**

Use this analogy to help understand depolarization and the impulse: Think of the positive ions as dominoes. When one positive ion enters the neuron, it causes the next ion to enter and then the next, and so on. When you push one domino over, it causes the next domino to fall, and so on. As you watch each domino fall, you see a wave-like activity. Figure 9-3 shows a depolarized neuron in sequence. To simplify this view, some of the dendrites have been removed.

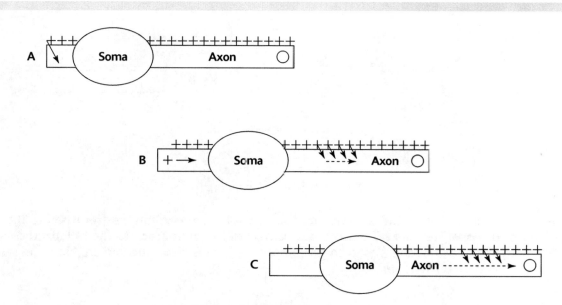

Figure 9-3: Depolarized neuron.

Figure 9-3a shows a polarized neuron. The neuron is stimulated and, therefore, one positive ion begins to enter the dendrite. This is the beginning of depolarization. This ion diffuses to the axon region. Figure 9-3b shows depolarization of the axon. The dotted arrow shows the movement of the impulse down the axon. Figure 9-3c shows more ions entering. The dotted arrow shows the impulse traveling a little farther down the axon. This process continues until the impulse reaches the end of the axon. At the end of the axon, there will be an influx of positive ions but in this case, it will be calcium ions. Calcium ions target the presynaptic vesicle and cause the vesicle to release the neurotransmitter into the synapse area. Figure 9-4 shows the release of the neurotransmitter.

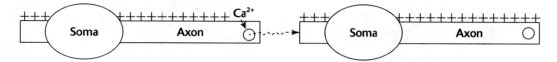

Figure 9-4: The release of the neurotransmitter.

As soon as the neurotransmitter (in this case, **acetylcholine**) is released from the presynaptic vesicle, it enters into the synaptic region. It flows across the synapse and comes in contact with the membrane of the next neuron in line. As soon as it stimulates the membrane of the next neuron, it causes depolarization of the next neuron in sequence, and the impulse continues on to its destination.

In order to use the neurons a second time, all the positive ions must leave the neuron and go back to the extracellular regions (outside of the neuron). Also, the neurotransmitter must be decomposed so it too leaves the synapse. Basically, everything needs to go back to its original position. The process of returning the ions to the ECF region is called **repolarization.** This is an active process and requires ATP.

The dendrite end releases an enzyme called **acetylcholinesterase.** This enzyme will decompose acetylcholine to form acetate and choline. Acetate and choline can then be reabsorbed into the presynaptic vesicle to recombine and make more acetylcholine. Figure 9-5 shows repolarization and the breakdown of the neurotransmitter.

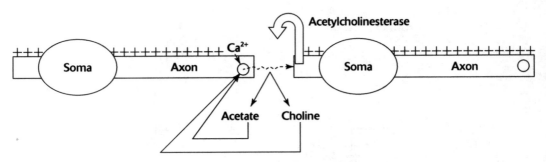

Figure 9-5: Repolarization.

After repolarization and after the breakdown of acetylcholine, everything has returned to the original state. The positive ions are back to the outside of the membrane to the polarized condition. The acetylcholine is decomposed so the synapse region is clear once again. Now the same set of neurons can be used again.

Example Problems

1. What are the three major parts of a neuron?

    answer: dendrite, soma, and axon

2. The process where by the positive ions are entering into the neuron is called

    _____.

    answer: depolarization

3. Acetylcholine is a neurotransmitter and _____ is an enzyme that breaks down acetylcholine.

    answer: acetylcholinesterase

4. Identify the direction that an impulse will travel through a neuron: _____, and then through the _____, and then through the _____.

    answer: dendrite, soma, and axon

5. What is the ion that is involved in causing the presynaptic vessel to release the neurotransmitter?

    answer: calcium ions (Ca^{2+})

Work Problems

1. The axon end (presynaptic vesicle) releases _____ and the dendrite end releases _____.

2. Give the name of a neurotransmitter and give the name of the enzyme that breaks down that neurotransmitter.

3. How many axons typically extend from the soma?

4. In order for a neuron to be used again, it must first _____.

5. Most of the positive ions associated with a neuron are _____ and _____ ions.

6. What initially causes positive ions to begin entering the neuron?

7. When a neuron repolarizes, it causes the neuron to return to the _____ state.

8. Upon depolarization, the positive ions enter into the (ICF or ECF)?

9. Upon release, neurotransmitters will stimulate what portion of the next neuron?

10. What is the name of the energy molecule required to get the positive ions back to the ECF area?

Worked Solutions

1. **a neurotransmitter;** an enzyme to break down the neurotransmitter

2. **acetylcholine;** acetylcholinesterase

3. **There can be several dendrites but generally just one axon.**

4. **repolarize**

5. **sodium ions and calcium ions (Na⁺ and Ca²⁺)**

6. **a stimulus**

7. **polarized**

8. **ICF, intracellular fluid region**

9. **the dendrite of the next neuron**

10. **ATP (adenosine triphosphate)**

One-Way Transmission

The nervous system is designed to ultimately protect the body from harm. To ensure this, it is imperative that the impulses travel one direction through a neuron to arrive at the correct destination. If impulses were allowed to travel in any direction within a neuron, chaos would result. An impulse travels in this manner:

1. The dendrite end is stimulated.

2. Depolarization occurs, and positive ions diffuse toward the axon region (specifically the axon hillock).

3. After the ions have arrived at the axon, the axon is depolarized, thus having numerous sodium ions entering the ICF in a domino fashion.

4. An impulse is traveling through the axon.

5. At the axon end, a neurotransmitter is released from the synaptic vesicle.

6. The neurotransmitter goes across the synapse and stimulates the dendrite of the next neuron.

7. The impulse continues traveling through the next neuron in the same manner.

Now, take a look at the scenario if the impulse were to travel the opposite direction:

1. Depolarization occurs in the axon.

2. The impulse travels from the axon through the soma.

3. The impulse travels on through the dendrite.

4. In order for the impulse to get across the synapse, there has to be a neurotransmitter.

5. The dendrite end does not have neurotransmitters.

6. The dendrite end only has enzymes to break down neurotransmitters.

7. The impulse cannot reach any destination by traveling from the axon, through the soma, and through the dendrite.

The impulse must travel toward the axon end (**presynaptic vesicle region**) in order to be able to successfully cross the synapse to continue its journey to its destination.

All-or-None Principle

This principle states that if a stimulus is strong enough to cause depolarization, the entire neuron will depolarize in succession. An impulse will travel the length of the neuron. If a stimulus is not strong enough to cause depolarization, an impulse will not even begin. An impulse will not travel part way through an axon, and then quit. It goes all the way or not at all.

Threshold

Threshold can be defined as the amount of stimuli required to cause depolarization. If a person has a high threshold, they will, therefore, require more stimuli in order to have depolarization and therefore in order to respond to something. A neuron with a low threshold will have depolarization with very little stimuli. Whenever a doctor gives a patient a drug that numbs the feeling, the drug actually raised the threshold of the neuron to the point where depolarization is not occurring. Without depolarization, an impulse will not occur. Without an impulse, the brain does not sense any pain.

Example Problems

1. What causes depolarization?

   answer: A stimulus of some sort.

2. Which end of the neurons consists of the synaptic vesicle?

    answer: The axon end.

3. Describe the direction an impulse must travel in order to reach it destination.

    answer: The impulse must travel through the axon toward the presynaptic vesicle.

4. An impulse cannot get across the synapse from the dendrite end because dendrites do not release _____.

    answer: neurotransmitters

5. A lot of stimuli is required for depolarization to occur if the patient has a (high or low) threshold level.

    answer: high

The Neuroglial Cells

There are a variety of neuroglial cells. These cells do not conduct impulses. They serve to protect the neurons. The most notable neuroglial (glial) cell is the **Schwann cells.** These cells wrap around the axon, thus protecting it. These cells form a myelin sheath around the axon. Figure 9-6 shows the arrangement of the Schwann cells around the axon.

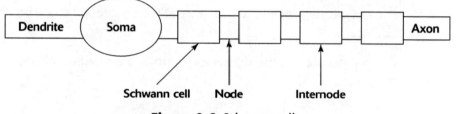

Figure 9-6: Schwann cells.

Figure 9-6 shows four Schwann cells wrapping around the axon. These cells form a **myelin sheath.** The node is the area of the axon that is not covered by a myelin sheath and is, therefore, unmyelinated. The Schwann cells form internodes. These are areas between the nodes. Because the internode areas are covered by Schwann cells, they are myelinated.

Nerves

Nerves are bundles of axons. **Axons** can be microscopic, such as those found associated with the brain, while some axons can be extremely long, such as the sciatic nerve. Typically when we think of a nerve, we think of the nerves associated with the peripheral nervous system. This is the system of nerves emerging from the brain or spinal cord and extending to the periphery of the body. The axons of many nerves are myelinated such as described in Figure 9-6.

Work Problems

1. The myelin sheath that is formed around the axon of some neurons is made of cells called
   _____.

2. If a dentist pokes on a patient's tooth, an impulse will travel to the brain and the brain
   will interpret pain. To prevent this interpretation, the dentist will give the patient some
   Novocain. Novocain prevents the brain from making any interpretation because Novocain
   will _____ the threshold level.

3. The nervous system is made of neurons and neuroglial cells. Schwann cells are what type
   of cell?

4. The Schwann cells typically surround and therefore protect the (axon, soma, dendrite).

5. The node associated with myelinated neurons is actually the exposed _____.

6. If the stimulus is below the threshold level, _____ will not occur.

7. Nerves are actually bundles of _____.

8. When an "impulse" is going across the synapse, it is traveling one direction only. It travels
   from the (axon end or dendrite end) _____ of one neuron to
   the (axon end or dendrite end) _____ of the next neuron in
   sequence.

9. According to the all-or-none principle, once depolarization occurs, there will be an
   _____.

10. Isolating and therefore protecting the nervous system is the function of what type of
    neural cells?

Worked Solutions

1. **Schwann cells**

2. **raise**

3. **neuroglial cells**

4. **axon**

5. **axon**

6. **depolarization**

7. **axons**

8. **axon end of the first neuron to the dendrite end of the next neuron**

9. **impulse**

10. **glial cells (neuroglial cells)**

Chapter Problems and Solutions

Problems

1.   Which part of the neuron is typically myelinated?

2.   The ions needed to get the neurotransmitter into the synaptic region are _____.

3.   The minimum amount of stimulus required to create depolarization is known as the _____.

4.   What is the principle that states that once depolarization has occurred, an impulse will travel to its destination?

5.   Diffusion of sodium ions occurs in the dendrites upon activation. Where does the action potential occur within the neuron?

6.   Most neurons lack centrioles. Therefore, neurons cannot undergo _____.

7.   Multiple sclerosis is a demyelinating disease. Therefore, multiple sclerosis is the destruction of the _____ cells that protect the neurons.

8.   Cell organelles are located in which part of a neuron? _____

9.   What is the name of the gap located between neurons or between a neuron and an organ?

10. In the synapse, the "impulse" cannot travel from the dendrite end to the axon of the next neuron because the dendrite does not release a _____.

Answers and Solutions

1.   **The axon is typically protected by the Schwann cells.** The Schwann cells from the myelin sheath.

2.   **calcium ions**

3.   **threshold level**

4.   **all-or-none principle**

5.   **in the axon**

6.   **Because they lack centrioles, they cannot produce spindle fibers.** Without spindle fibers, they cannot undergo cell reproduction.

7.   **neuroglial cells or Schwann cells**

8.   **Cell organelles are located in the soma.**

9.   **synapse**

10. **neurotransmitter**

Supplemental Chapter Problems

Problems

1. The cells involved in processing information are called _____, while the cells involved in protecting those cells are called _____.

2. The cells involved in transferring information to different parts of the body are called _____.

3. The space between each neuron is called a _____.

4. Novocain creates "numbness" in the nerves by preventing the positive ions from entering the neuron. Therefore, Novocain inhibits the action called _____.

5. True or false: A drug that inhibits depolarization could cause paralysis. _____

6. Insecticides inhibit the action of acetylcholinesterase. Therefore, _____ cannot be decomposed in the synapse thereby resulting in nerve malfunction.

7. Nicotine is a drug that mimics acetylcholine and therefore could act as a _____ and cause depolarization of the neurons.

8. Caffeine is a drug that can cause some people to develop the "shakes" by having unwanted muscle contraction. This is because caffeine causes the neuron membrane to become very permeable to positive ions. This means that caffeine makes depolarization (easier or more difficult) _____ to occur.

9. Tetradotoxin is a seafood toxin that prevents sodium ions from entering the neuron. Therefore, the toxin inhibits the action called _____.

10. The process of reestablishing the sodium ions in the ECF region of the neuron is called _____.

Answers

1. neurons; glial

2. neurons

3. synapse

4. depolarization

5. true. Without depolarization, an impulse will not occur.

6. acetylcholine

7. stimulus

8. easier

9. depolarization

10. repolarization

Chapter 10
The Central Nervous System

The entire nervous system is divided into two major divisions. One is the central nervous system (CNS) and the other is the peripheral nervous system (PNS). The central nervous system is comprised of the brain and spinal cord, while the peripheral nervous system is comprised of the nerves that go to and from the CNS. The peripheral nerves are associated with the periphery of the body. This chapter discusses the brain and the spinal cord.

The Brain

When we first study the brain, we typically view the external aspects of the brain. Table 10-1 lists a few select features of the external view with a brief description of the location for each.

Table 10-1 External Brain	
Brain Region	**Location**
Frontal lobe	This area of the brain is associated with the frontal bone of the skull.
Occipital lobe	This area of the brain is associated with the occipital bone of the skull.
Parietal lobe	This area of the brain is associated with the parietal bone of the skull. It is located between the frontal lobe and the occipital lobe.
Temporal lobe	This area of the brain is associated with the temporal bone of the skull.
Limbic lobe	This is the only lobe of the brain that is located on the inside of the brain.
Longitudinal fissure	This is the deep groove that goes the length of the brain. It separates the left hemisphere with the right hemisphere and is therefore a sagittal fissure.
Lateral cerebral fissure	This is a deep groove that divides the temporal lobe with the rest of the brain. It is on the lateral side of the brain.
Sulcus	These are the little grooves in the brain. The plural form of the word is sulci.
Gyrus	These are the little ridges found on the brain. The plural form of the word is gyri.
Cerebellum	This structure is located mainly inferior to the occipital lobe of the brain.
Pons	As you examine the brainstem, the first big bulge you encounter is the pons.
Medulla oblongata	Inferior to the pons is a smaller bulge called the medulla oblongata.

Figure 10-1 shows the external view of the brain and has letters pointing to the various structures identified in Table 10-1. Figure 10-2 shows the internal, sagittal view of the brain and has letters identifying the various lobes identified in Table 10-1.

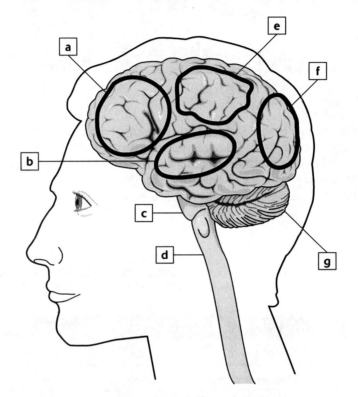

Figure 10-1: External view of the brain.

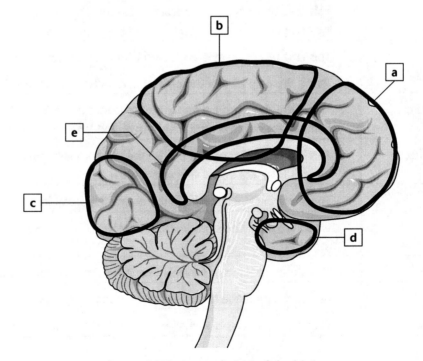

Figure 10-2: Internal view of the brain.

Example Problems

Answer the following questions in reference to Figure 10-1 and 10-2 and Table 10-1.

1. Based on the description in Table 10-1, what is the name of brain area a on Figure 10-1 and 10-2?

   answer: frontal lobe

2. Based on the description in Table 10-1, what is the name of brain area b on Figure 10-1 and area d on Figure 10-2?

   answer: temporal lobe

3. Based on the description in Table 10-1, what is the name of brain structure c?

   answer: pons

4. Based on the description in Table 10-1, what is the name of brain structure d?

   answer: medulla oblongata

5. Based on the description in Table 10-1, what is the name of brain area e on Figure 10-1 and area b on Figure 10-2?

   answer: parietal lobe

6. Based on the description in Table 10-1, what is the name of brain area f on Figure 10-1 and area c on Figure 10-2?

   answer: occipital lobe

7. Based on the description in Table 10-1, what is the name of brain structure g?

   answer: cerebellum

8. What is the name of area e on Figure 10-2?

   answer: limbic lobe

The Brain's Five Major Divisions

The brain has been divided into five major divisions. Each division is made of a variety of parts. Table 10-2 lists the five divisions and a few select parts associated with each division.

Table 10-2 Brain Divisions and Associated Structures	
Brain Division	**Select Parts Associated with the Division**
Telencephalon	The cerebrum is the main part of this division. The cerebrum consists of the various lobes discussed previously.
Diencephalon	This division consists of the thalamus and hypothalamus. The hypothalamus is located slightly inferior and anterior to the thalamus.

(continued)

Table 10-2 Brain Divisions and Associated Structures *(continued)*	
Brain Division	**Select Parts Associated with the Division**
Mesencephalon	This division consists of the midbrain region. This region is located between the thalamus and the pons.
Metencephalon	This division consists of the cerebellum and the pons. The cerebellum is inferior to the occipital lobe and is posterior to the pons.
Myelencephalon	This division consists of the medulla oblongata. The medulla oblongata is the small bulge located inferior to the pons.

Figure 10-3 is a sagittal view of the brain. The brain divisions are identified by numbers. The numbered leader lines are pointing to the various structures discussed in Table 10-2.

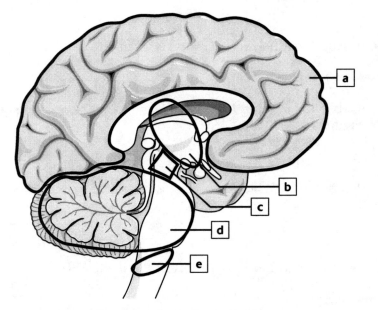

Figure 10-3: The brain divisions.

Example Problems

Answer the following questions in reference to Figure 10-3 and Table 10-2.

1. Based on the description in Table 10-3, what is the name of the brain division identified by area a?

   answer: telencephalon

2. Based on the description in Table 10-3, what is the name of the brain division identified by area b?

   answer: diencephalon

3.   Based on the description in Table 10-3, what is the name of the brain division identified by area c?

     answer: mesencephalon

4.   Based on the description in Table 10-3, what is the name of the brain division identified by area d?

     answer: metencephalon

5.   Based on the description in Table 10-3, what is the name of the brain division identified by area e?

     answer: myelencephalon

Other Structures of the Brain

Other structures of the brain that can be seen from a sagittal view (see Figure 10-4) are identified and described in Table 10-3.

Table 10-3 Sagittal View of the Brain	
Brain Structure	*Brief Description*
Cerebrum	This is the major part of the brain. It consists of folds (gyri) and valleys (sulci).
Corpus callosum	This is curved, elongated white matter that connects the two hemispheres together.
Fornix	This structure is a group of nerves that appear to extend inferiorly to the corpus callosum but is not connected to the corpus callosum.
Lateral ventricle	This ventricle is a cavity that is filled with fluid. It is located between the corpus callosum and the fornix.
Choroid plexus	This is a group of cells that are located just on the inferior border of the fornix. These cells produce cerebrospinal fluid.
Pituitary gland	This is a small gland that is located inferior and connected to the hypothalamus.
Limbic lobe (limbic system)	This lobe is an elongated gyrus that is located immediately superior to the corpus callosum.
Thalamus	This is a general area that is located inferior to the corpus callosum and the fornix.
Hypothalamus	This is a general area that is located slightly inferior and slightly anterior to the thalamus.
Pituitary gland	The pituitary has a stalk that attaches it to the inferior portion of the hypothalamus.
Pons	As you descend the brain stem, you encounter a rather large bulge. This first large bulge is the pons.
Medulla oblongata	Inferior to the pons is a smaller bulge called the medulla oblongata.
Midbrain	Located between the pons and thalamus region.
Cerebellum	This structure is located inferior to the occipital lobe of the brain.

Figure 10-4 shows the sagittal view of the brain and has numbers on the various structures identified in Table 10-3.

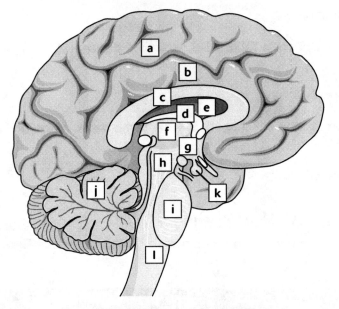

Figure 10-4: Sagittal view of the brain.

Example Problems
Answer the following questions in reference to Figure 10-4.

1. Based on the description in Table 10-4, what is the name of the brain part identified by the area a?

    answer: cerebrum

2. Based on the description in Table 10-3, what is the name of the brain part identified by area b?

    answer: limbic lobe

3. Based on the description in Table 10-3, what is the name of the brain part identified by the area c?

    answer: corpus callosum

4. Based on the description in Table 10-3, what is the name of the brain part identified by the area d?

    answer: fornix

5. Based on the description in Table 10-3, what is the name of the brain part identified by the area e?

    answer: lateral ventricle

6. Based on the description in Table 10-3, what is the name of the brain part identified by the area f?

   answer: thalamus

7. Based on the description in Table 10-3, what is the name of the brain part identified by the area g?

   answer: hypothalamus

8. Based on the description in Table 10-3, what is the name of the brain part identified by the area h?

   answer: midbrain

9. Based on the description in Table 10-3, what is the name of the brain part identified by the area i?

   answer: pons

10. Based on the description in Table 10-3, what is the name of the brain part identified by the area j?

    answer: cerebellum

11. Based on the description in Table 10-3, what is the name of the brain part identified by the area k?

    answer: pituitary gland

12. Based on the description in Table 10-3, what is the name of the brain part identified by the area l?

    answer: medulla oblongata

Table 10-4 lists a few select brain structures or regions and describes a brief function for that part or region.

Table 10-4 Functions of the Various Brain Structures	
Brain Structure	**Brief Function**
Frontal lobe	This portion of the brain controls many of the fine motor movements.
Occipital lobe	This lobe is involved in the interpretation of vision.
Temporal lobe	This lobe is involved in interpreting odors and hearing.
Parietal lobe	This lobe is involved with the interpretation of all senses except hearing, smelling, and vision.
Limbic lobe	This lobe is involved in long-term memory.

(continued)

Table 10-4 Functions of the Various Brain Structures *(continued)*	
Brain Structure	**Brief Function**
Corpus callosum	This group of nerves connects the two hemispheres together. It also provides communication between the two hemispheres.
Medulla oblongata	This structure controls the rhythmic pattern of the heart rate and breathing. This structure also causes you to cough or sneeze in an effort to expel possible irritants, thereby protecting the lungs.
Pons	This structure alters the rhythmic pattern of the heart rate and breathing.
Cerebellum	Instructions from the cerebrum tell the body to start walking. This information is sent to the cerebellum, and the cerebellum actually carries out the task of walking. The cerebellum ensures that the body moves in a coordinated manner.
Midbrain	This region controls muscle tone. This region also contains nerves referred to as the reticular activating system (RAS). This system keeps us alert.
Thalamus	This structure acts as a relay station, thereby taking external information and sending it to various parts of the cerebrum for storage.
Hypothalamus	This structure consists of nerves that are involved in detection of thirst. It also is involved with various behavior characteristics.

Work Problems

1. What is the name of the structure that connects the right hemisphere with the left hemisphere?

2. The pons is located _____ (use a directional term) to the medulla oblongata.

3. Identify the structure that carries out the functions of the cerebrum in a nice, smooth, coordinated manner.

4. Identify the lobe that is involved in interpreting vision.

5. The infundibulum is the name of the stalk that connects the pituitary gland to the _____.

6. Which lobe is responsible for long term memory?

7. Which lobe allows us to pick up a pencil and use our fine motor skills?

8. The limbic lobe is part of which brain division?

9. The _____ is a cluster of cells that will produce the cerebrospinal fluid.

10. A rather large fluid filled cavity of the brain is called the _____.

Worked Solutions

1. **corpus callosum**

2. **superior**

3. **cerebellum**

4. **occipital lobe**

5. **hypothalamus**

6. **limbic lobe**

7. **frontal lobe**

8. **telencephalon**

9. **choroid plexus**

10. **lateral ventricle**

The Spinal Cord

The spinal cord serves as a major pathway for impulses to travel to the brain for the interpretation of data and also away from the brain to cause response. The pathways are typically referred to as tracts (ascending tracts and descending tracts). The spinal cord also collects some information and relays it directly to some body organ or tissue to create a response. In other words, the brain is not always involved.

The following bullet list outlines some aspects of the spinal cord in its entirety.

❑ The spinal cord extends from the brain (medulla oblongata region) to the first lumbar vertebrae.

❑ The distal end comes to a point (close to lumbar vertebrae number one) called the **conus medullaris.**

❑ From the conus medullaris region, the spinal cord forms numerous branches called **cauda equina** (because it looks like a horse's tail).

❑ There is one lone filament that extends into the coccyx region called the **filum terminale.**

❑ There are 31 pairs of nerves that branch off the spinal cord.

❑ There are 8 cervical spinal nerves, 12 thoracic spinal nerves, 5 lumbar spinal nerves, 5 sacral spinal nerves, and 1 coccygeal nerve.

❑ There are 8 cervical spinal nerves but only 7 cervical vertebrae. Spinal nerve number 1 emerges from between the skull and cervical vertebrae 1.

❑ The 5 sacral nerves pass through the foramen on the sacrum.

❑ In a transverse view, gray and white matter can be seen. The white matter surrounds the gray matter.

❑ The axons in the spinal cord give the white matter its name. It is the axons that form the tracts that ascend or descend the spinal cord.

❏ The ascending tracts transmit sensory information to the brain.

❏ The descending tracts transmit motor information away from the brain.

Figure 10-5 shows the spinal nerves branching off the spinal cord. Only the right side is shown.

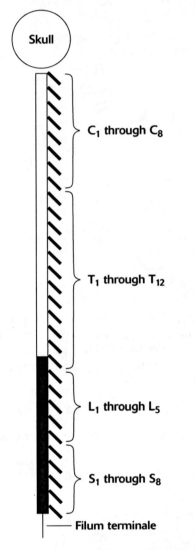

Figure 10-5: The spinal cord and spinal nerves.

Nerve Plexuses

The nerves branching off the spinal cord can branch some more and also merge together. As the nerves merge together, they form a braided network. This braided network of nerves is called a **plexus.** Below is a select list of nerves to be discussed.

❏ There is a plexus of nerves associated with the cervical nerves. One major plexus emerging from the cervical region eventually forms a single nerve that innervates the diaphragm muscle. This nerve is called the **phrenic nerve.**

❑ Another plexus arises from the lower cervical and upper thoracic region. This is called the **brachial plexus.** This plexus will eventually form a single nerve that innervates the medial side of the arm. If you bump this nerve (on the medial side of the elbow region), you will feel a tingling sensation on your arm and hand. This is what some people refer to as "hitting your funny bone." This nerve is called the **ulnar nerve.**

❑ Another plexus arises from the lumbar region. This is called the **lumbar plexus.** One major nerve that develops from this plexus controls some of the muscles of the upper thigh. This nerve is called the **femoral nerve.**

❑ The last plexus arises from the sacral region. This is called the **sacral plexus.** One major nerve that develops from this plexus is the **sciatic nerve.** This nerve controls some of the posterior thigh muscles and foot muscles.

❑ Notice, the thoracic region does not form a plexus. All of the nerves emerging from the thoracic region form single nerves, not plexuses.

Example Problems

1. How many pairs of nerves emerge from the spinal cord?

   answer: 31 pair

2. How many coccygeal nerves are there?

   answer: just one, the filum terminale

3. How many different plexuses are there?

   answer: four: cervical, brachial, lumbar, and sacral

4. What is the name of one region of the spinal cord that does not form a plexus?

   answer: thoracic

5. The plexus that forms a nerve that controls the major breathing muscle is called the

   answer: cervical plexus

Transverse View of the Spinal Cord

Examination of the transverse view of the spinal cord helps to understand the various tracts running through the spinal cord. Figure 10-6 shows a transverse view of the spinal cord and points out the white matter and the gray matter. The white matter consists of numerous axons and the gray matter consists of numerous soma and dendrites. Table 10-6 lists a few select facts about the white and gray matter of the spinal cord. The best way to orient the transverse view of the spinal cord is to understand that the anterior side of the spinal cord has a rather wide fissure.

Table 10-5 Transverse Spinal Cord	
Structure	**Brief Function and Location**
Posterior white	This area consists of columns of nerves that transmit information to the brain for the interpretation of fine touch and pressure. These are **ascending** tracts.
Lateral white	This area consists of columns of nerves that transmit information to the cerebellum for the interpretation of balance. These are **ascending** tracts. This area also consists of tracts of nerves that transmit information from the cerebrum and brainstem for precise limb movement and digit movement, respectively. These are **descending** tracts.
Anterior white	This area consists of columns of nerves that transmit information to the thalamus region (pain and temperature information). The thalamus then relays the information to the cerebrum. These are **ascending** tracts. This area also consists of tracts of nerves that transmit information from the brainstem to the limbs for balance. These are **descending** tracts.
Posterior gray	This area consists of sensory nerves that come from the periphery of the body. It also consists of sensory nerves that come from the internal organs.
Lateral gray	This area consists of motor nerves that go to the internal organs.
Anterior gray	This area consists of motor nerves that go to the periphery of the body.

Figure 10-6 shows the transverse view of the spinal cord with arrows pointing the various regions described in Table 10-5.

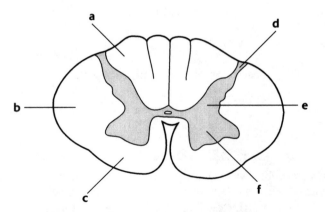

Figure 10-6: Transverse spinal cord.

Example Problems

For Questions 1 through 3, identify which region is white matter and which region is gray matter.

1. Region a consists of _____.

    answer: white matter

2. Region e consists of _____.

   answer: gray matter

3. Region f consists of _____.

   answer: gray matter

For Questions 4 through 6, identify which region correlates with the described function or area.

4. Which region transmits information to the cerebellum?

   answer: region b

5. Which region picks up information from internal organs and sends information to the cerebrum for interpretation?

   answer: region d

6. Which region is considered to be the anterior white?

   answer: region c

Protecting the Central Nervous System

The brain and the spinal cord are protected by three layers of membrane tissue and bone tissue. The membrane that is nearest the brain and spinal tissue is the **pia mater.** The next membrane is called the **arachnoid.** The membrane nearest the skull or vertebrae is the **dura mater.**

Flowing between the pia mater and the arachnoid is the **cerebrospinal fluid.** Flowing between the arachnoid and the dura mater is **venous fluid** (blood).

The CSF (cerebrospinal fluid) is produced by the choroid plexus and enters into the subarachnoid space. From there, the CSF flows through arachnoid villi to enter into the subdural space.

Therefore, the brain and spinal cord are protected by membranes, fluid, and bone.

The combination of all three membranes is called the **meninges.**

Work Problems

1. The (anterior or posterior) side of the spinal cord has a rather deep fissure.

2. The conus medullaris is located approximately the area of which vertebrae?

3. The sensory tracts of the spinal cord are (ascending or descending) tracts.

4. Many of the nerves that branch off the spinal cord and extend through the arm came from which plexus?

5. The three membranes that serve to protect the brain and spinal cord are collectively called _____.

6. Which type of nerves cause action (motor nerves or sensory nerves)?

7. Cerebrospinal fluid will enter into the blood by absorbing through what structures associated with the arachnoid membrane?

8. The posterior white region of the spinal cord consists of (ascending or descending tracts).

9. The gray matter of the spinal cord is made of (soma or axons).

10. There are _____ cervical vertebrae and _____ cervical nerves.

Worked Solutions

1. **anterior**

2. **lumbar 1 (L₁)**

3. **ascending**

4. **brachial plexus**

5. **meninges**

6. **motor nerves**

7. **arachnoid villi**

8. **ascending tracts because that region consists of sensory nerves.**

9. **soma.** Axons make up the white matter.

10. **seven; eight**

Chapter Problems and Solutions

Problems

1. A defect in the _____ may make it difficult for a person to recognize whether they are thirsty or not.

2. The limbic lobe consists of a major gyrus called the cingulate gyrus. This gyrus is located superior to what brain structure?

3. What major fissure is located between the temporal lobe and the frontal lobe (refer to Figure 10-1)?

4. What is the name of the structure that comprises the metencephalon and is located between the pons and the thalamus regions?

5. The cerebrospinal fluid serves as a partial protection for the brain and spinal cord. It is produced by the _____.

6. Hitting this part of the skull and therefore stimulating this lobe of the brain could cause a person to literally "see" stars. What is that part?

7. The cerebellum works closely with which brain division?

8. The lateral ventricle is a fluid filled ventricle that is located between the corpus callosum and the _____.

9. What is the name of the structure that causes us to cough or sneeze in an effort to expel possible harmful substances, thereby providing protection for the lungs?

10. Figure 10-4 is a picture of which hemisphere of the brain?

11. A person may not be able to feel fine touch sensations if which part of the spinal cord was damaged?

12. The posterior gray matter consists of _____ (sensory or motor) nerves.

13. What is the name of the plexus that is associated with nerves of the arm?

14. The ascending tracts of the spinal cord are transmitting information (to or from) the cerebrum.

15. The nerve impulses that go away from the spinal cord are traveling through (motor or sensory) nerves and the impulses that go toward the spinal cord are traveling through (motor or sensory) nerves.

Answers and Solutions

1. **hypothalamus**

2. **corpus callosum**

3. **lateral cerebral fissure**

4. **midbrain**

5. **choroid plexus**

6. **occipital lobe**

7. **telencephalon.** The cerebrum is part of the telencephalon.

8. **diencephalon.** The thalamus is part of the diencephalon.

9. **medulla oblongata**

10. **left**

11. **posterior white**

12. **sensory**

13. **brachial plexus**

14. **to**

15. **motor and sensory**

Supplemental Chapter Problems

Problems

1. The spinal cord proper itself extends only to the conus medullaris. This area is located in the vicinity of which vertebrae?

2. The area of an infant's skull that is commonly referred to as the "baby's soft spot" is anatomically known as the anterior fontanels. This is the area where the skull has not fused together yet and, therefore, you can actually feel one of the membranes associated with the meninges. Which membrane of the meninges can be felt?

3. Cerebrospinal fluid flows between which two meningeal membranes?

4. A patient has a viral infection that has destroyed the anterior gray matter of the spinal cord in the lumbar region. Which part of the body will probably be most affected?

5. Which consists of only sensory tracts; the white matter or gray matter of the spinal cord?

6. Alzheimer's disease affects the part of the brain that is involved in long term memory. Therefore, Alzheimer's disease affects which specific part (lobe) of the brain?

7. Viral meningitis results in fewer fatalities than does bacterial meningitis. Regardless, however, both types of meningitis affects the membranes that cover the brain and spinal cord collectively called the _____.

8. Cerebrospinal fluid enters the venous flow (blood circulation) by passing through structures between the arachnoid and dura mater called the _____.

9. What is the name of the structure that is located inferior and slightly anterior to the thalamus region?

10. In order to walk from point A to point B, the cerebrum initiates the impulse. This impulse is then sent to the _____, which will then send the impulse on to the leg muscles to create smooth, coordinated movement.

Answers

1. lumbar 1 (L$_1$)

2. dura mater

3. pia mater and arachnoid

4. This person will have difficulty walking because the anterior gray matter has nerves that control the leg muscles. These nerves emerge from the lumbar region.

5. white matter

6. limbic lobe

7. meninges

8. arachnoid villi

9. hypothalamus

10. cerebellum

Chapter 11
The Peripheral Nervous System

The peripheral nervous system (PNS) consists of nerves that are associated with some part of the periphery of the body (as opposed to the central nervous system, CNS, which consists of the brain and spinal cord). The peripheral nervous system is further divided into afferent nerves and efferent nerves. The afferent nerves are nerves that transmit information *to* the CNS, and the efferent nerves are nerves that transmit information *away from* the CNS. The efferent nerves can be divided into somatic nerves and autonomic nerves. The **somatic nerves** send information to the skeletal muscles. The **autonomic nerves** send information to the internal organs. The autonomic nerves can be divided even further yet. One division consists of the sympathetic nerves and the other consists of the parasympathetic division. Typically, the **sympathetic nerves** increase metabolism and alertness. Antagonistic to that are the **parasympathetic nerves,** which typically conserve energy. Figure 11-1 shows a flow chart that illustrates the various subdivisions of the peripheral nervous system.

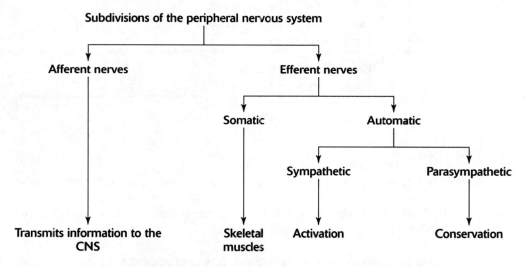

Figure 11-1: The anatomy of the peripheral nervous system.

The Spinal Nerves

The spinal nerves were introduced in Chapter 10. The spinal nerves are peripheral nerves as they branch off the spinal cord to go to various parts of the body. Be sure to study the spinal nerves and nerve plexuses within Chapter 10. The nerves associated with the sympathetic nerves emerge from the thoracic region of the spinal cord. The nerves associated with the parasympathetic nerves emerge from the brain and the sacral region of the spinal cord. Table 11-1 gives a brief overview of the function of the sympathetic nerves. Table 11-2 gives a brief overview of the function of the parasympathetic nerves.

Table 11-1 Sympathetic Nerves	
Emerges from:	*Select Function*
Upper thoracic region	Dilates the pupils of the eyes Dilates respiratory tubes Increases heart rate
Middle thoracic region	Increases adrenalin production from the adrenal glands
Lower thoracic region	Relaxes urinary bladder Causes ejaculation

Table 11-2 Parasympathetic Nerves	
Emerges from:	*Select Function*
The brain	Constricts the pupils of the eye Constricts the respiratory tubes Decreases heart rate
The sacral region	Tenses the urinary bladder Causes erection

Example Problems

1. Which system (sympathetic or parasympathetic) allows a person to breathe better?

    answer: sympathetic. It causes the respiratory tubes to dilate thus allow for better air exchange.

2. Generally, if some part of our body is activated, it is probably due to impulses coming from the (sympathetic or parasympathetic) nerves.

    answer: sympathetic

3. The dilation of the pupils comes from nerves associated with the _____ and the constriction of the pupils comes from nerves associated with the

    _____.

    answer: upper thoracic region of the spinal cord; brain

4. The _____ nerves transmit impulses to the brain and/or spinal cord and the _____ nerves transmit impulses away from the brain and/or spinal cord.

    answer: afferent; efferent

5. The _____ nerves of the efferent division control the visceral organs of the body.

    answer: autonomic

The Cranial Nerves

The cranial nerves are 12 pairs of nerves that emerge from the brain and are part of the peripheral nervous system because they transmit information from the CNS to the periphery of the body or from the periphery of the body to the CNS. The cranial nerves that transmit information from the CNS to the periphery are called **motor nerves.** The cranial nerves that transmit information from the periphery to the CNS are called **sensory nerves.** Some of the cranial nerves can transmit to and from the CNS and they are called **mixed nerves.**

The cranial nerves are associated with the inferior portion of the brain. They are numbered (with Roman numerals) in sequence as they are positioned anterior to posterior on the brain.

Table 11-3 lists the cranial nerves and their function.

Table 11-3 The Cranial Nerves			
Number	**Nerve Name**	**Function**	
I	Olfactory	Sensory	Transmits information for the interpretation of smell.
II	Optic	Sensory	Transmits information for the interpretation of vision.
III	Oculomotor	Motor	Transmits information to control some eye muscles. Transmits information to control the pupil dilation and the shape of the lens.
IV	Trochlear	Motor	Transmits information to control some eye muscles.
V	Trigeminal	Mixed	Consists of three major branches: **Ophthalmic:** Transmits information from the forehead region. **Maxillary:** Transmits information from the upper lip region. **Mandibular:** Transmits information to the muscles of the jaw for chewing.
VI	Abducens	Motor	Transmits information to control some eye muscles.
VII	Facial	Mixed	Transmits information to control face muscles and to produce tears. Transmits information for the interpretation of taste.
VIII	Vestibulocochlear	Sensory	Consists of two major parts: **Vestibular:** Transmits information for the interpretation of balance. **Cochlear:** Transmits information for the interpretation of hearing.

(continued)

Table 11-3		The Cranial Nerves *(continued)*	
Number	**Nerve Name**	**Function**	
IX	Glossopharyngeal	Mixed	Transmits information for the interpretation of taste and pain sensations from the tongue. Transmits information to control swallowing.
X	Vagus	Mixed	Transmits information for the interpretation of visceral organ activity. Transmits information to control the visceral organs such as slow down the heart rate and breathing rate.
XI	Spinal accessory	Motor	Transmits information to control the sternocleidomastoid and trapezius muscle.
XII	Hypoglossal	Motor	Transmits information to move the tongue.

Examine the following Figures to help understand the information listed in Table 11-3 and also to understand the significance of learning the skull foramen. Figure 11-2 illustrates the sensory cranial nerves. Remember, the sensory nerves transmit information to the CNS, hence the arrows going toward the CNS. In order for the nerves to reach the CNS, they have to pass through a foramen. Figure 11-3 illustrates the motor cranial nerves, and Figure 11-4 illustrates the mixed cranial nerves.

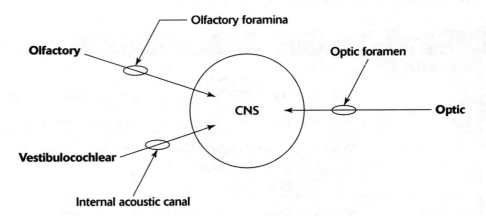

Figure 11-2: Sensory cranial nerves.

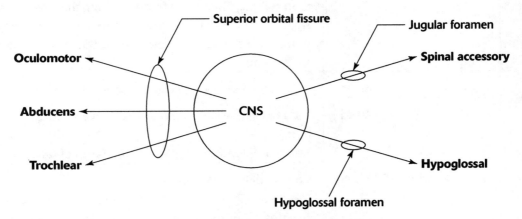

Figure 11-3: Motor cranial nerves.

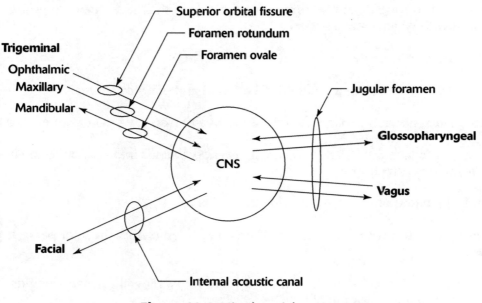

Figure 11-4: Mixed cranial nerves.

Example Problems

1. A patient is having difficulty in moving his tongue. A pinched cranial nerve could cause this problem. Which nerve would this be?

   answer: hypoglossal nerve

2. A damaged abducens nerve could cause problems doing what activity?

   answer: movement of the eye

3. One cranial nerve is involved in taste sensations from the tongue and another cranial nerve is involved in pain sensations from the tongue. Identify those two cranial nerves.

   answer: The facial nerve is involved with taste sensations and the glossopharyngeal nerve is involved with pain sensations from the tongue.

4. How many cranial nerves have something to do with some aspect of the eye?

   answer: five: optic, oculomotor, trochlear, abducens, and facial.

5. Cranial nerve III is located (anterior or posterior) to the optic nerve?

   answer: posterior. The cranial nerves are numbered in sequence from anterior to posterior.

Work Problems

1. How many cranial nerves are sensory, motor, and mixed?

2. How many cranial nerves pass through the jugular foramen?

3. If a patient is having difficulty controlling one side of their face, they are probably having a problem with which cranial nerve?

4. How many cranial nerves are strictly afferent nerves?

5. How many cranial nerves are combined afferent and efferent nerves?

6. Which cranial nerve is probably involved when a person feels pain from a sinus headache?

7. If we begin to lose our balance, which cranial nerve sends information to the brain so we can make the corrections?

8. What is the name of the most anterior cranial nerve?

9. Nerves associated with T_3 and T_4 would be considered (sympathetic or parasympathetic) nerves.

10. When food is pushed to the back of the throat, the swallowing process begins. This is largely due to the activation of which cranial nerve?

Worked Solutions

1. **sensory = 3, motor = 5, mixed = 4**

2. **three**

3. **facial nerve**

4. **3: olfactory, optic, and vestibulocochlear (specifically, the vestibular portion).** Afferent is the same as sensory.

5. **4: trigeminal, facial, glossopharyngeal, and vagus.** These are the mixed nerves.

6. **trigeminal nerve, specifically the ophthalmic portion**

7. **vestibulocochlear**

8. **olfactory.** Cranial nerves are numbered in sequence from anterior to posterior.

9. **sympathetic.** The parasympathetic nerves emerge from the brain or sacral region.

10. **glossopharyngeal**

Chapter Problems and Solutions

Problems

1. Name the cranial nerves that pass through the jugular foramen.

2. Identify the cranial nerve that is involved in the production of tears.

3.   In order to tilt our head from side to side, we must contract the sternocleidomastoid muscle. Which cranial nerve is involved in that process?

4.   What is the function of the abducens and trochlear nerves?

5.   How many **different** foramen (or other openings in the skull) do the cranial nerves pass through?

6.   Which part of cranial nerve V consists of mainly motor nerves?

7.   The activation of the (sympathetic or parasympathetic) nerves typically creates the "fight or flight" response.

8.   A patient can no longer develop an erection. When questioning the patient, the physician discovered that the patient had recently fallen extremely hard on his "tail bone." In this case, the (sympathetic or parasympathetic) nerves were probably damaged.

9.   Several cranial nerves pass through the superior orbital fissure. Where is this fissure located?

10. How many cranial nerves send information to various muscles of the body?

Answers and Solutions

1.   **spinal accessory, glossopharyngeal, and vagus**

2.   **facial**

3.   **spinal accessory**

4.   **They control some of the eye muscles.**

5.   **8: olfactory foramina, optic foramen, internal acoustic canal, jugular foramen, superior orbital fissure, hypoglossal foramen, foramen ovale, and foramen rotundum.**

6.   **mandibular**

7.   **sympathetic.** These nerves typically speed up metabolism and the heart rate, and so on.

8.   **parasympathetic**

9.   **The superior orbital fissure is a fissure (slit) located at the back of the eye socket region.**

10. **8: abducens (eye muscles), trochlear (eye muscles), oculomotor (eye muscles), trigeminal (jaw muscle), facial (face muscles), glossopharyngeal (swallowing muscles), spinal accessory (neck and shoulder muscles), and hypoglossal (tongue muscle).**

Supplemental Chapter Problems

Problems

1. When a patient goes into a vagal response, their blood pressure and heart rate drop drastically. This is because the vagus nerve is a (sympathetic or parasympathetic) nerve.

2. When you bite your tongue, you feel tremendous pain. Which cranial nerve was stimulated in order for you to feel the pain?

3. In order to focus on objects near and far, the shape of the lens has to change for proper focus. Which cranial nerve is involved in that process?

4. Some people take drugs to enhance their athletic ability, which of course, is not fair. Which type of drug (sympathetic or parasympathetic) would be the "drug of choice" in this situation?

5. A person walks into a well-lit room. A physician notices that the pupils of his eyes are constantly dilated (in a well-lit room they should be constricted). The physician might suspect that this person is taking a (sympathetic or parasympathetic) drug.

6. Bell's palsy is a viral condition that usually affects one of the cranial nerves. It typically causes patients to lose control of one side of their face. Bell's palsy is a condition that affects which cranial nerve?

7. The process of olfaction is the function of which cranial nerve?

8. The process of gustation is the function of which cranial nerve?

9. An increase in the release of adrenalin will give a person an extra burst of energy. However, damage to which portion of the spinal cord has an affect on the release of adrenalin?

10. One of the cranial nerves branches in the region of the zygomatic bone. One branch extends to the maxillary region, one branch extends to the forehead region, and the third branch extends to the masseter muscle. Which cranial nerve is being described?

Answers

1. parasympathetic nerve

2. glossopharyngeal

3. oculomotor

4. sympathetic

5. sympathetic

6. facial nerve (VII)

7. olfactory (olfaction is the process of smelling)

8. facial (gustation is the process of tasting)

9. thoracic region (specifically, the middle thoracic)

10. trigeminal nerve

Chapter 12

The Five Senses

The various senses of the body can be divided into two main groups. One is the **general senses,** which consist of the interpretation of pain, touch, temperature, and pressure. The other group of senses is **special senses,** which consist of the interpretation of vision, hearing, tasting, smelling, and balance.

Introducing the Five Senses

Table 12-1 lists the five special senses and a few select facts about each one.

Table 12-1 The Five Senses	
Special Sense	**Select Function**
Smell	The process of smelling is called **olfaction.**
	The nerves associated with olfaction pass through the olfactory foramina.
	Olfactory nerves are the only cranial nerves that go directly to the cerebrum and do not have to be relayed by the thalamus.
	These nerves are identified as cranial nerve I.
Taste	The process of tasting is called **gustation.**
	The cells involved are called gustatory cells.
	These cells are clustered inside taste buds.
	Taste buds line the papillae of the tongue.
	The papillae are the bumps on the surface of the tongue.
	Dissolved food particles stimulate the gustatory cells and a signal is sent via the facial nerve (cranial nerve VII) to the brain for interpretation.
Vision	Light enters the pupil of the eye and stimulates special cells that make up the retina of the eye.
	Once these cells are stimulated, they send an impulse along the optic nerve and then the impulse is relayed to the occipital lobe of the brain.
	The special cells of the retina are called rods and cones.
	The rods function during dim light conditions. They do not sense color.
	The cones function only during bright light conditions. They sense color.

(continued)

Table 12-1 **The Five Senses** (*continued*)	
Special Sense	**Select Function**
Hearing	Sound waves create vibrations in the cochlear region of the ear. These vibrations stimulate special cells that send an impulse via the cochlear nerve (of the vestibulocochlear nerve, cranial nerve VIII) to the brain for interpretation.
Balance	The movement of the body causes movement of fluid within the semicircular canals of the internal ear. This fluid movement stimulates special cells that send an impulse via the vestibular nerve (of the vestibulocochlear nerve, cranial nerve VIII) to the brain for interpretation.

Example Problems

1. What structure in the ear is responsible for hearing? _____

   answer: the cochlea

2. What structure in the ear is responsible for balance? _____

   answer: the vestibular apparatus

3. Cranial nerve (I, II, VII, or VIII) goes directly to the cerebrum? _____

   answer: I. This is the olfactory nerve and is the only one that goes directly to the cerebrum.

4. True or false: The bumps on the surface of the tongue are the taste buds?

   answer: false. The taste buds are located inside the bumps (papillae).

5. The retina of the eye is made of cells called _____ and _____.

   answer: rods and cones

The Tongue

As food enters the mouth and is partially dissolved, the dissolved food stimulates the gustatory cells of the taste buds. These cells send a signal (via the facial nerve) to the brain for the interpretation of taste. Figure 12-1 shows a magnified view of the tongue to show the relationship of the gustatory cells, taste buds, and papillae.

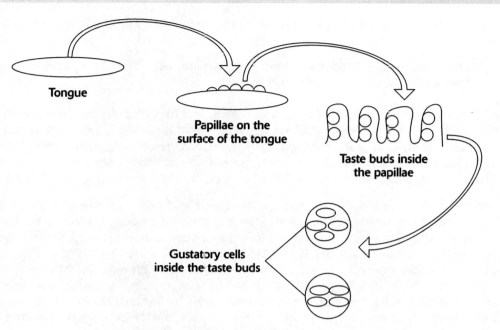

Figure 12-1: The tongue.

The Pathway of Light into the Eyes

Light rays pass through the pupil, and ultimately to the cells of the retina at the back of the eye. The retina cells send signals to the brain (via the optic nerve) for the interpretation of vision. Figure 12-2 shows the internal components of the eye.

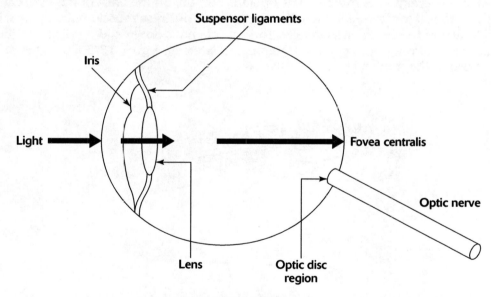

Figure 12-2: Internal eye.

1. Light passes through the cornea of the eye.

2. Light passes through the pupil.

3. Light passes through the lens.

4. The lens will focus the light on the retina of the eye.

5. In order for the light to reach the retina, it has to pass through the vitreous chamber of the eye.

6. The light will activate the rods and cones of the retina, which will send a nerve signal via the optic nerve to the occipital lobe of the brain.

The retina consists of a combination of **rods** and **cones.** The **fovea centralis** is the area of the retina that consists of 100% cones. Therefore, the best vision is during the day when you are look-ing straight at an object. During the daylight, only the cones are operating. The rest of the retina has a combination of rods and cones; therefore, there aren't as many cones as there are in the fovea centralis region. This is why peripheral vision is not as clear as looking straight at an object.

The optic disc region is the area where the blood vessels and optic nerve exit the eye. In this location, there aren't any rods or cones. If an image is focused directly on the optic disc, vision will not be interpreted. Because the optic nerve exits the eye a little bit medial to center, it is not in the same location for both eyes. So, if an object is focused on the optic disc in the right eye, that same object will be focused elsewhere on the retina of the left eye. So, basically, whatever the right eye can't "see," the left eye can. When we look at objects, rarely are our eyes station-ary. Therefore, if a person has only one eye, objects won't be focused on the optic disc unless the eye is kept very still. Because of the lack of vision in the optic disc region, this area is also known as the blind spot.

The lens can change shape in order to properly focus an image on the retina. The lens changes its shape due to the contraction of the ciliary muscles attached to the suspensor ligaments.

There are six muscles that control the movement of the eye. The **lateral rectus** muscle of the eye rotates the eye laterally. The **medial rectus** muscle rotates the eye medially. The **superior rectus** muscle rotates the front of the eye upward. The **inferior rectus** muscle rotates the front of the eye downward. The **superior oblique** muscle pulls the eye in such a manner to allow you to see down and laterally. This muscle pulls up and medially on the back of the eye, causing the front of the eye to rotate down and laterally. The **inferior oblique** muscle pulls the eye in such a manner to allow you to see up and laterally. This muscle pulls down and medially on the back of the eye, causing the front of the eye to rotate up and laterally. Figure 12-3 illustrates the six muscles of the eye (the medial rectus is not shown).

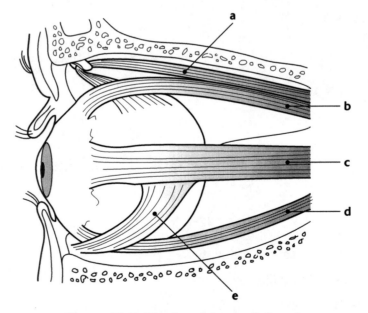

Figure 12-3: Muscles of the eye (left eye).

Example Problems

Questions 1 through 5 can be answered by studying the information pertaining to Figure 12-3.

1. What is the name of the muscle represented by letter a? _____

   answer: superior oblique

2. What is the name of the muscle represented by letter b? _____

   answer: superior rectus

3. What is the name of the muscle represented by letter c? _____

   answer: lateral rectus

4. What is the name of the muscle represented by letter d? _____

   answer: inferior rectus

5. What is the name of the muscle represented by letter e? _____

   answer: This is a portion of the inferior oblique. The rest of this muscle goes posterior, just like the superior oblique.

Work Problems

1. What is the name of the area of the retina that doesn't consist of any rods?

2. What is the name of the area of the retina that doesn't consist of any rods or cones?

3. The large fluid filled region located between the lens and the retina is called the _____.

4. The opening that is created by the movement of the iris structure, which allows light to enter the eye is called the _____.

5. Which retinal cells are actively involved in daylight vision? _____

6. The contraction of which eye muscle will cause the left eye to rotate laterally?

7. The contraction of which eye muscle will cause the eye to rotate downward so as to "look" at the floor?

8. The bumps on the surface of the tongue are not taste buds. The taste buds are located inside the bumps. The bumps are called _____.

9. When the gustatory cells are activated, they will send signals to the brain for interpretation via the _____ nerve (cranial nerve name).

10. Most of the special sensory nerves have to be relayed by the _____ before "reaching" the cerebrum for interpretation.

Worked Solutions

1. **The fovea centralis contains 100% cones.**

2. **The optic disc or blind spot is devoid of rods or cones due to the presence of the optic nerve and blood vessels.**

3. **vitreous chamber**

4. **pupil**

5. **cones**

6. **lateral rectus**

7. **inferior rectus**

8. **papillae**

9. **facial (CN VII)**

10. **thalamus**

The Pathway of Sound Waves into the Ear

The inner ear consists of the hearing apparatus and the balance apparatus. The hearing apparatus is the cochlea, which is a snail-shaped organ. Sound waves vibrate the ossicles, which, in turn, causes the movement of fluid inside the cochlea. This fluid movement stimulates a special organ inside the cochlea called the **organ of Corti.** The organ of Corti stimulates cranial nerve VIII, which transmits a signal to the brain for the interpretation of hearing.

While looking at Figure 12-4, consider the pathway of sound waves into the ear:

1. Sound waves are trapped by the helix (pinna) of the ear.

2. Sound waves travel through the ear canal (auditory canal).

3. The sound waves will cause the tympanic membrane to vibrate.

4. This vibration causes the first ossicle (malleus) to vibrate.

5. The vibrating malleus causes the second ossicle (incus) to also vibrate.

6. The vibrating incus will cause the third ossicle (stapes) to vibrate.

7. The vibrating stapes will cause the fluid inside the cochlea to begin moving. The stapes covers an area called the oval window of the cochlea.

8. The movement of the fluid stimulates the cells, within the cochlea, which in turn stimulates the cochlear nerve (a portion of the vestibulocochlear nerve, CN VIII).

9. A signal is sent to the brain for the interpretation of hearing.

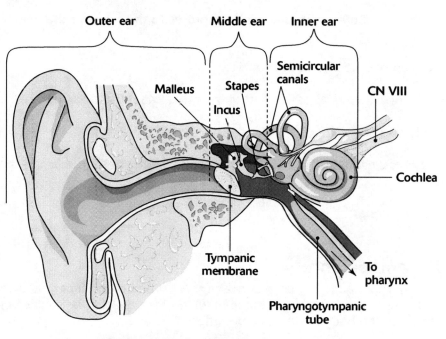

Figure 12-4: The pathway of sound waves.

Consider the following additional information regarding the ear:

❑ The outer ear is comprised of the auricle (helix, antihelix, tragus, and antitragus) and the auditory canal. The **helix** is the outer rim, the **antihelix** is the inner rim, the **tragus** is the part you push on to block hearing, and the **antitragus** is the bump on the antihelix opposite the tragus.

❑ The middle ear is comprised of the tympanic membrane, the ossicles, and the pharyngotympanic tube (formerly known as the Eustachian tube).

The pharyngotympanic tube is often called the **auditory tube** (not to be confused with the auditory canal, which is a canal that extends from the helix to the tympanic membrane). The pharyngotympanic tube extends from the ossicle region to the pharynx region.

❑ The inner ear is comprised of the cochlea and the vestibular apparatus.

Example Problems

Questions 1 through 5 can be answered by studying the information pertaining to Figure 12-4.

1. Which ossicle is attached to the tympanic membrane?

   answer: malleus

2. Which ossicle is attached to the cochlea?

   answer: stapes

3. What is the name of the organ that sends impulses to the brain for the interpretation of hearing?

   answer: Organ of Corti within the cochlea

4. The malleus, incus, and stapes are part of the outer ear, middle ear, or inner ear?

   answer: middle ear

5. Which part of the ear actually "collects" the sound waves to send them to the tympanic membrane?

   answer: helix (pinna)

The Inner Ear and Balance

The ear is involved in more activity than just hearing. A portion of the inner ear is involved with balance. When a person moves his head, the fluid within the semicircular canals begins to move, thus sending signals to the brain for the interpretation of balance.

- ❏ The inner ear is comprised of the cochlea and the vestibular apparatus.
- ❏ The cochlea is responsible for hearing. The vestibular apparatus is responsible for balance.
- ❏ The vestibular apparatus is comprised of the vestibule and the three semicircular canals.
- ❏ When the body moves, this sets up fluid motion inside the semicircular canals.
- ❏ The movement of this fluid stimulates cells, which in turn stimulate the vestibular portion of the vestibulocochlear nerve (CN VIII).
- ❏ The signal is then interpreted in the brain regarding acceleration and balance.

The Middle Ear and Pressure

If you are in an airplane as it ascends or descends, you may experience ear pain. This is due to the fact that at high altitudes, air pressure is less than compared to ground level. This means that the pressure in the middle ear area (medial side of the tympanic membrane) is ground-level pressure and the pressure in the ear canal (lateral side of the tympanic membrane) is less-than-ground-level pressure. This causes the tympanic membrane to bulge outward, and this bulging causes pain. To equalize the pressure, many people chew gum. The action of chewing gum causes movement of the pharyngotympanic tube, which, in turn, helps to move air out of the middle ear region, thus equalizing the pressure.

- ❏ The middle ear consists of the **ossicles, tympanic membrane,** and the **pharyngotympanic tube.**
- ❏ The pharyngotympanic tube is formerly known as the Eustachian tube. Eustachian is not a descriptive term. It is a word based on a person's name. Pharyngotympanic is a descriptive term. It is a tube that extends from the tympanic region to the pharynx.
- ❏ This tube equalizes the pressure inside the middle ear with outside pressure.
- ❏ If this tube is blocked, pressure begins to build in the middle ear, which in turn puts pressure on the tympanic membrane, resulting in pain.

❏ Children are prone to earaches. Sometimes, lymphatic tissue can grow to the point of blocking the opening to the pharyngotympanic tube. This blockage creates a build-up of pressure within the middle ear. As time progresses, the pharyngotympanic tube will secrete mucus, which blocks the tube even more. More pain will result.

❏ A bacterial infection can also cause problems in the middle ear. The bacteria can cause the pharyngotympanic tube to produce excess mucus, thus resulting in blockage of the tube.

Work Problems

1. Identify the ossicles in sequence from the tympanic membrane to the oval window.

2. Part of the vestibulocochlear nerve is associated with balance while the other part is associated with _____.

3. Each vestibular apparatus consists of _____ (how many?) semicircular canals?

4. Cranial nerve number _____ is the vestibulocochlear nerve.

5. The vestibulocochlear nerve emerges from the cochlea and passes through the _____ foramen (canal) to arrive at the brain for interpretation.

6. Hearing occurs when the cells within the _____ of the internal ear are stimulated by the vibrating motion of fluid.

7. A viral infection of the _____ ear can cause a person to become dizzy. (Use one of these terms; outer, middle, and inner).

8. The sense of equilibrium is detected in which portion of the ear? Use one of these terms: outer, middle, and inner. _____

9. When a person doesn't want to hear something, they push on a structure of their outer ear to block the sound waves. What is the name of this structure?

10. The pharyngotympanic tube is a tube that serves to equalize pressure within the _____ ear region and the outside.

Worked Solutions

1. **malleus, incus, stapes**

2. **hearing**

3. **3**

4. **VIII**

5. **internal acoustic**

6. **cochlea**

7. **inner (this area consists of the vestibular apparatus for balance)**

8. **inner**

9. **tragus**

10. **middle**

Chapter Problems and Solutions

Problems

1. The contraction of the _____ muscle of the left eye and the contraction of the _____ muscle of the right eye will cause the eyes to become cross-eyed.

2. The contraction of the _____ muscle of the left eye and the contraction of the _____ muscle of the right eye will cause the eyes to rotate to the right.

3. What is the name of the transparent outer covering of the eye? _____

4. Each eye is controlled by (how many) extrinsic eye muscles? _____

5. The _____ causes the oval window of the cochlea to vibrate.

6. Where are gustatory cells located? _____

7. The _____ is the continuation of the sclera (the white portion of the eye), thus forming a transparent layer on the anterior side of the eye.

8. The central opening of the eye that allows light to pass through is the _____.

9. The lens is able to focus objects on the retina due to the action of which muscles?

10. When we look straight at an object we can see it clearly and with the sharpest color. This is because the image is focusing on the _____, which consists of 100% cones.

Answers and Solutions

1. **medial rectus of the left eye and medial rectus of the right eye**

2. **medial rectus of the left eye and lateral rectus of the right eye**

3. **cornea**

4. **6:** lateral rectus, medial rectus, superior rectus, inferior rectus, superior oblique, and inferior oblique

5. **stapes**

6. **within the taste buds**

7. **cornea**

8. **pupil**

9. **ciliary muscles**

10. **fovea centralis**

Supplemental Chapter Problems

Problems

1. The organ of Corti is an organ in the ear that is responsible for hearing. The organ of Corti is located in what internal ear structure?

2. The special senses are made of the following types of special cells: gustatory cells, olfactory cells, and cone cells. Which of those types of cells are involved in odors?

3. If a person has a great deal of difficulty seeing at night, it could be the fact that they have diminished functioning _____ cells of the eye.

4. A person all of a sudden drives into a blinding snow storm. They cannot see any distance at all due to the snow. They begin to feel as though the car is beginning to go to the right when in fact the car is still going straight. A person gets this feeling because the movement of the car causes movement of the fluid in the _____. The movement of the fluid causes the brain to interpret imbalance (movement to the right in this scenario).

5. Dogs have a much greater sense of smell compared to humans. This is because dogs have far more _____ cells (receptors) as compared to humans.

6. A middle ear infection could hinder the movement of the _____ and therefore making it difficult for the patient to hear.

7. The sclera is the white part of the eye. As the sclera curves to form the anterior portion of the eye, it differentiates and becomes transparent. The transparent portion is called the _____.

8. An infection of the inner ear could affect a patient's hearing and also his or her _____.

9. When a person spins around several times, they feel dizzy and feel like they are still moving. This is because the fluid in the _____ is still moving even though the person is stationary.

10. Having the ability to see distant objects, but near objects are fuzzy is called _____ and having the ability to see near objects, but far objects appear fuzzy, is called _____ (use these terms to fill in the blanks; hyperopia and myopia).

Answers

1. cochlea

2. olfactory

3. rod

4. semicircular canals

5. olfactory

6. ossicles

7. cornea

8. balance

9. semicircular canals

10. hyperopia; myopia

Chapter 13
The Endocrine System

The nervous system allows the body to respond to various stimuli in a quick manner, and this allows for homeostasis. The endocrine system (with its array of hormones) also allows the body to respond to various stimuli. This endocrine response, however, is typically slower than the neural response, yet, in spite of this slower response, it is still necessary in order to maintain homeostasis.

The best way to study the endocrine system is to study it in this manner:

- ❑ Identify the gland or tissue that is producing the hormone.
- ❑ Identify the hormone that is produced.
- ❑ Identify the target for that hormone (where is that hormone going in the body?).
- ❑ Identify the function of that hormone (what is that hormone going to cause the body to do?).

This chapter is organized according to that study guide.

The Pituitary Gland

Table 13-1 lists the hormones that are released by the pituitary gland. Table 13-2 identifies the target for that hormone. Table 13-3 identifies the function for that hormone.

Table 13-1 Pituitary Hormones	
Pituitary region	*Hormone*
Posterior	Antidiuretic hormone (ADH)
Posterior	Oxytocin (OT)
Anterior	Growth hormone (somatotropin) (GH)
Anterior	Prolactin (PRL)
Anterior	Thyroid stimulating hormone (TSH)
Anterior	Adrenocorticotropic hormone (ACTH)
Anterior	Follicle stimulating hormone (FSH)
Anterior	Luteinizing hormone (LH)

Table 13-2 identifies the target for the hormones released by the pituitary gland. When a hormone targets a specific tissue or organ, it causes that tissue or organ to respond in a specific manner.

Table 13-2 Target of Pituitary Hormones	
Hormone	**Hormone Target**
Antidiuretic hormone (ADH)	Internal kidney tubules called nephrons
Oxytocin (OT)	Uterus and mammary tissue
Growth hormone (somatotropin) (GH)	General body cells
Prolactin (PRL)	Mammary tissue
Thyroid Stimulating Hormone (TSH)	Thyroid gland
Adrenocorticotropic Hormone (ACTH)	Cortex of the adrenal gland
Follicle Stimulating Hormone (FSH)	In males, this hormone targets internal tubules of the testes called seminiferous tubules. In females, this hormone targets the ovaries.
Luteinizing Hormone (LH)	In males, this hormone targets special cells inside the testes called interstitial cells. In females, this hormone targets the follicles inside the ovaries.

Table 13-3 identifies the function of the pituitary hormones. These hormones will cause specific tissues or organs to respond in a specific manner, which is designed to help maintain homeostasis.

Table 13-3 Function of Pituitary Hormones	
Hormone	**Hormone Function**
Antidiuretic hormone (ADH)	Prevents the kidney tubules from losing water thereby preventing dehydration.
Oxytocin (OT)	Causes the uterus to contract during labor. Causes the mammary tissue to release milk during nursing.
Growth hormone (somatotropin) (GH)	Causes protein synthesis for growth.
Prolactin (PRL)	Causes the mammary tissue to produce milk.
Thyroid stimulating hormone (TSH)	Causes the thyroid gland to release hormones such as; calcitonin, thyroxine, and triiodothyronine.
Adrenocorticotropic hormone (ACTH)	Causes the adrenal cortex to release a group of hormones called glucocorticoids.
Follicle stimulating hormone (FSH)	In males, this hormone causes the seminiferous tubules to produce sperm cells. In females, this hormone causes the maturation of an egg.

Hormone	Hormone Function
Luteinizing hormone (LH)	In males, this hormone causes the interstitial cells to release testosterone. Testosterone causes the development of secondary sex characteristics such as: facial hair and deeper voice. In females, this hormone causes the follicles, which contain a mature egg, to release the egg (ovulation). It also causes the ruptured follicle to release progesterone and estrogen.

Example Problems

1. There are two hormones (oxytocin and prolactin) that target the mammary tissues and are associated with milk. Identify the hormone that causes milk production.

   answer: prolactin. Oxytocin causes the release of milk, not the production of milk.

2. Cortisone is a hormone that is part of a group of hormones called glucocorticoids. Based on this, what structure releases cortisone?

   answer: adrenal cortex

3. The _____ hormone causes the maturation of an egg and the _____ hormone causes the ovulation of an egg.

   answer: follicle stimulating; luteinizing

4. The thyroid gland produces and releases some hormones. What is the name of the hormone that causes the thyroid gland to release its hormones?

   answer: thyroid stimulating hormone

5. The antidiuretic hormone prevents water loss by the kidneys by targeting what structures of the kidneys?

   answer: nephrons of the kidneys

Other Endocrine Glands

The pituitary gland is considered to be the "master gland" because it releases some hormones that control the action of other glands of the body. Even though the other glands are not master glands, they still have profound effects on the body. Table 13-4 lists a few select glands and the hormones they release. Table 13-5 lists the targets for those hormones, and Table 13-6 identifies the function of those hormones.

Table 13-4 Other Glands and Hormones	
Gland	**Hormone released**
Thyroid	Calcitonin, thyroxine (T_4), and triiodothyronine (T_3)
Parathyroid	Parathormone (parathyroid hormone or PTH)
Thymus	Thymosin
Atria of the heart	Atrial natriuretic peptide (ANP)
Pancreas	Insulin and glucagon
Adrenal medulla	Epinephrine (adrenalin) and norepinephrine (noradrenalin)
Adrenal cortex	Glucocorticoids and aldosterone
Kidney	Erythropoietin (EPO)

Table 13-5 identifies the target for certain hormones. When a hormone targets a specific tissue or organ, it causes that tissue or organ to respond in a specific manner.

Table 13-5 Target for Select Hormones	
Hormone	**Hormone Target**
Calcitonin	Bone cells (osteoblasts)
Thyroxine and triiodothyronine	General body cells
Parathormone (PTH)	Bone cells (osteoclasts)
Thymosin	Lymphocytes (a specific white blood cell)
Atrial natriuretic peptide (ANP)	Tubules (nephrons) inside the kidneys
Insulin	Cell membrane protein channels
Glucagon	Liver cells
Adrenalin and noradrenalin (Epinephrine and norepinephrine)	General body cells
Glucocorticoids	General body cells
Aldosterone	Nephrons of the kidneys
Erythropoietin (EPO)	Bone marrow (specifically the erythroblasts within the bone marrow).

Table 13-6 identifies the function for certain hormones. After the hormone targets its specific tissue, it causes the tissue to respond in a specific way, which leads to homeostasis.

Table 13-6 Function for Select Hormones	
Hormone	*Hormone Function*
Calcitonin	Causes the osteoblasts to remove calcium ions from the bloodstream for the purpose of making bone material. This decreases blood calcium ion levels.
Thyroxine and triiodothyronine	Increases cellular metabolism.
Parathormone (PTH)	Causes the osteoclasts to remove calcium ions from the bones. Calcium ions then enter into the bloodstream. This increases blood calcium ion levels.
Thymosin	Converts lymphocytes to T cells involved in the immune system.
Atrial natriuretic peptide (ANP)	Causes the nephrons to remove water and sodium ions from the bloodstream and send them to the urinary bladder to be voided.
Insulin	Causes protein channels within the membrane to form pores thereby letting glucose enter into the cell for metabolism. This decreases the blood glucose levels.
Glucagon	Causes the liver cells to break down glycogen to form glucose. The glucose then enters into the bloodstream. This increases the blood glucose levels.
Adrenalin and noradrenalin (Epinephrine and norepinephrine)	Causes cells to increase their metabolic rates. Causes the mobilization of glucose into the bloodstream to be used by cells.
Glucocorticoids	Causes cells to increase their metabolic rates. Causes an increase in glucose production from non-carbohydrate products (**gluconeogenesis).**
Aldosterone	Causes the nephrons of the kidneys to put water and sodium ions back into the bloodstream. This results in the conservation of water and sodium ions.
Erythropoietin (EPO)	Causes the erythroblasts to begin the formation of erythrocytes (red blood cells).

Example Problems

1. The kidneys play a vital role in homeostasis. According to Tables 13-1 through 13-6, how many hormones target some aspect of the kidneys?

   answer: three: aldosterone, ANP, ADH

2. According to Tables 13-1 through 13-6, how many hormones speed up metabolic processes?

   answer: five: thyroxine, triiodothyronine, epinephrine, norepinephrine, and glucocorticoids

3. According to Tables 13-1 through 13-6, how many hormones target some aspect of bone?

    answer: three: erythropoietin, calcitonin, and parathormone

4. According to Tables 13-1 through 13-6, how many hormones are involved in controlling sodium ion levels?

    answer: two: aldosterone and atrial natriuretic peptide

5. According to Tables 13-1 through 13-6, how many hormones are involved in controlling calcium ion levels?

    answer: two: calcitonin and parathormone

Work Problems

1. Insulin lowers blood glucose levels. Name several hormones that raises blood glucose levels.

2. Aldosterone causes the kidneys to conserve sodium ions. Name the hormone that causes the kidneys to lose sodium ions.

3. Parathormone will raise blood calcium ion levels. Name the hormone that lowers blood calcium ion levels.

4. The organ or tissue that responds to the presence of a hormone is called the hormone's _____.

5. Triiodothyronine (T_3) is made of three iodine atoms. How many atoms make up thyroxine? _____

6. The thymus gland is located _____ to the heart. (Use one directional term for your answer).

7. The parathyroid glands are located _____ to the thyroid gland. (Use one directional term for your answer).

8. What causes the release of testosterone?

9. White blood cells help the body fight infections. Therefore, the white blood cells are part of the immune system. What hormone functions as part of the immune system?

10. The outer layer of the adrenal gland produces aldosterone. This layer of the adrenal gland is called the adrenal _____.

Worked Solutions

1. **glucagon, epinephrine, norepinephrine and glucocorticoids (for example)**

2. **atrial natriuretic peptide (ANP)**

3. **calcitonin**

4. **target**

5. **4 (T$_4$)**

6. **superior**

7. **posterior**

8. **lutenizing hormone**

9. **thymosin (thymosin targets lymphocytes)**

10. **cortex**

Chapter Problems and Solutions

Problems

1. Any hormone that causes the body to retain water (puts water into the bloodstream) could cause the blood pressure to rise. Identify two hormones that could cause an increase in blood pressure.

2. Antagonistic hormones are hormones that work the opposite of each other. Therefore, the main antagonistic hormone to insulin would be _____.

3. Which portion of the pituitary releases the most hormones? _____

4. Identify the hormone that causes egg development in females and sperm development in males.

5. The adrenal medulla produces _____ and _____, and the adrenal cortex produces _____ and _____.

6. What hormone triggers the release of glucocorticoids?

7. Glucagon can cause the formation of glucose from glycogen (a carbohydrate) in the liver. Glucocorticoids can cause the formation of glucose from non-carbohydrates such as amino acids. This process is called _____.

8. Which gland lies within the abdominopelvic cavity near the stomach and small intestine?

9. Diabetes insipidus is a condition where the patient lacks the production of antidiuretic hormone. Diabetes mellitus is a condition where the patient lacks the ability to produce _____.

10. A patient experiencing a decreased production of thyroxine and triiodothyronine could be due to the lack of which hormone coming from the pituitary gland?

Answers and Solutions

1. **aldosterone and antidiuretic hormone**

2. **glucagon**

3. **anterior pituitary**

4. **follicle stimulating hormone**

5. **epinephrine and norepinephrine; aldosterone and glucocorticoids**

6. **adrenocorticotropic hormone (ACTH)**

7. **gluconeogenesis**

8. **pancreas**

9. **insulin**

10. **thyroid stimulating hormone**

Supplemental Chapter Problems

Problems

1. Complete removal of the thyroid gland also results in the removal of which gland?

2. A decrease in parathormone could cause a decrease in what ion in the bloodstream?

3. If a patient had a lack of sodium ions, their body would begin to produce what hormone?

4. The antidiuretic hormone is produced by the hypothalamus of the brain but is released by the _____.

5. If a patient has a high concentration of aldosterone in their bloodstream, they probably also have a high concentration of what ion in their bloodstream?

6. An increase in aldosterone could cause the blood pressure to go (up or down).

7. When blood glucose levels rise in a patient, which hormone will be released to compensate the rise of glucose?

8. A patient has an increased number of red blood cells. This could be due to the result of an increased amount of which hormone?

9. Gonadotropins are a group of hormones that regulate the male and female reproductive organs. Identify two gonadotropins previously discussed.

10. Increased amount of blood calcium ions would result in an increased amount of _____ (identify a hormone) released in order to get the blood calcium ion levels back to a normal level.

Answers

1. parathyroid

2. calcium ions

3. aldosterone

4. posterior pituitary gland

5. sodium ions

6. up

7. insulin

8. erythropoietin (EPO)

9. follicle stimulating hormone and luteinizing hormone

10. calcitonin

Chapter 14
The Blood

This chapter focuses on the circulating blood. The circulating blood provides a vital function for survival. The blood transports hormones to their destination, it transports waste products to the kidneys for removal, and it transports ions to various parts of the body.

Blood Composition

If a sample of blood were put into a test tube and prevented from clotting, it would separate into two major layers. The surface layer would account for about 55%, and the lower layer would be about 45%. The surface layer is **plasma,** which is mostly water, and the bottom layer is **cellular,** of which about 99% would be erythrocytes and about 1% would be leukocytes. The cellular layer (bottom layer) would further separate into erythrocytes and leukocytes. The **erythrocytes** are denser than leukocytes so they become the most inferior layer, and the **leukocytes** will settle between the erythrocyte layer and plasma. Figure 14-1 illustrates the various layers of blood and Table 14-1 lists the components and specific facts about each component.

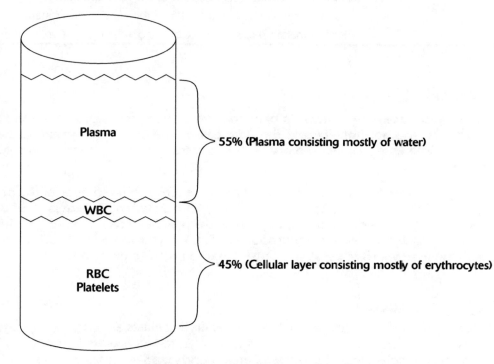

Figure 14-1: The layers of whole blood.

Table 14-1 Facts About Whole Blood Components	
Whole Blood Component	*Brief Facts*
Plasma	Constitutes about 55% of the whole blood.
	Consists of 92% water.
	Consists of 7% protein.
	Consists of 1% ions and minerals.
	Plasma consists of 3 major proteins.
	Fibrinogen: Involved with blood clotting.
	Albumin: Contributes to blood osmotic pressure.
	Globulins: Serve as antibodies/bind to and transports hormones.
	Two major ions in the plasma: sodium ions (Na^{1+}) and chloride ions (Cl^{1-}).
	Other ions are potassium ions (K^{1+}) and calcium ions (Ca^{2+}).
Cellular	Constitutes about 45% of the whole blood.
	Consists of three major cellular components.
	Erythrocytes: Transport oxygen and carbon dioxide. There are about 5 million per cubic mm of blood.
	Leukocytes: Defend the body against pathogens. There are about 5,000 to 11,000 per cubic mm of blood.
	Platelets: Involved in blood clotting. There are about 350,000 to 450,000 per cubic mm of blood.
	Erythrocytes make up about 99% of the cellular component.
	Leukocytes make up less than 1% of the cellular component.
	Platelets make up less than 1% of the cellular component.

Red Blood Cells

Red blood cells (**erythrocytes**) are formed in bone marrow. A hormone (**erythropoietin**) is produced by the kidneys. This hormone targets the stem cells of the bone marrow and causes the cells to differentiate into erythrocytes. The process of converting stem cells into erythrocytes is called **erythropoiesis.**

The function of erythrocytes is to transport oxygen from the lungs to the tissues of the body and to transport carbon dioxide from the tissues to the lungs for exhalation. If there is a deficiency in erythrocytes, there will be a deficiency in oxygen. A decrease in oxygen arriving at the kidneys will trigger the kidneys to release erythropoietin (EPO). Erythropoietin will target the **erythroblasts** (bone marrow stem cells) and cause them to undergo erythropoiesis.

Erythrocyte Structure

Erythroblasts have a nucleus and cell organelles. As the erythroblasts are undergoing erythropoiesis, they lose their nucleus and cell organelles and begin to incorporate hemoglobin in their place. **Hemoglobin** is a molecule that is made of amino acids (and is therefore considered to be a protein molecule) and heme. **Heme** is a complex organic molecule that consists of iron.

Oxygen binds to the iron portion of heme while carbon dioxide binds to the amino acid portion of the hemoglobin molecule.

Erythrocyte Life Span

During erythropoiesis, the erythroblasts lose their nucleus and cell organelles. Therefore, the anucleated red blood cells do not have anything to keep them alive. So, the erythrocyte has a very limited life span. Their life span is about 120 days. Because they don't have a nucleus or cell organelles, they cannot repair themselves (undergo mitotic events) if they are damaged.

Example Problems

1. Whole blood appears red because about 99% of the cellular components are
   _____

   answer: erythrocytes (red blood cells)

2. Erythrocyte is the name for red blood cells. What is the name for white blood cells?

   answer: leukocytes

4. Erythropoietin is produced and released by the _____ when _____ concentrations are low.

   answer: kidneys; oxygen

5. White blood cells are more dense than plasma but less dense than _____.

   answer: erythrocytes

6. The stem cells that are responsible for forming erythrocytes are called _____

   answer: erythroblasts

White Blood Cells

White blood cells (**leukocytes**) are formed from stem cells in bone marrow just as erythrocytes are. There are a variety of hormones and triggering mechanisms involved. The process of forming leukocytes is called **leukopoiesis.** There are five major types of leukocytes. Consider the following list of leukocytes and their functions.

❑ **Neutrophils:**

   About 50–70% of the leukocytes are neutrophils.

   Specialize in attacking and destroying bacteria.

❑ **Eosinophils:**

   About 2–4% of the leukocytes are eosinophils.

   These cells respond during an allergic reaction and also during a parasite infection such as worm infestation.

❑ **Basophils:**

   Less than 1% of the leukocytes are basophils.

   These cells release histamine that promotes inflammation that is involved in the healing of a wound.

❑ **Monocytes:**

About 2–8% of the leukocytes are monocytes.

Two monocytes can fuse together to produce a gigantic cell. This gigantic cell can engulf (phagocytize) rather large foreign particles. This gigantic cell is typically referred to as a **macrophage.**

❑ **Lymphocytes:**

About 20–40% of the leukocytes are lymphocytes.

Lymphocytes respond to bacterial infections as well as viral infections.

Leukocyte Response

If the body encounters an infection of some sort, the leukocytes will respond by undergoing cell reproduction and, thereby, increasing in numbers. For example, if the body experiences a bacterial infection, the neutrophils will respond by increasing in numbers in an effort to phagocytize the bacteria, thereby resulting in getting the body back into homeostasis. If the body experiences a viral infection, typically the lymphocytes will increase in numbers in an effort to destroy the viruses. Lymphocytes generally do not phagocytize; rather, they release a host of chemicals that will destroy viruses.

Examination of the leukocytes can help a physician determine why a patient is out of homeostasis. When a patient is in homeostasis, their leukocyte count would be 5,000 to 11,000 WBCs/mm^3. A laboratory technologist can examine a sample of blood and actually count the number of WBCs a patient has. For example, if the count is in excess of 11,000/mm^3, the laboratory technologist will then do a **differential count.** A differential count indicates the number of each cell type out of a random sample of 100 cells. A percentage can easily be determined and is then compared to the normal values such as those in the preceding section. If the laboratory technologist determines that the percentage of neutrophils is high, that's a good indication that the patient has a bacterial infection. The technologist can then culture the bacteria to determine what kind it is and based on that report, the physician can then prescribe a specific type of antibiotic that is designed to kill that specific kind of bacteria.

Example Problems

1. Which leukocyte releases histamine? _____

    answer: basophil

2. If a patient has a viral infection, which leukocyte would you expect to be high in numbers?

    answer: lymphocytes: other leukocytes may increase too but lymphocytes are the main ones regarding viruses.

3. Which leukocyte is the least common in a patient who is in homeostasis? _____

    answer: basophil

4. A laboratory technologist did a differential count of the leukocytes and determined there to be 11 lymphocytes out of 100 cells counted. Is this value too high or too low regarding normal lymphocyte counts?

   answer: this would be too low. 11 lymphocytes out of 100 = 11% (11/100 × 100 = 11%). A normal count is 20–40%.

5. A patient who is allergic to ragweed probably has a higher than normal number of _____. (Choose one of the 5 leukocytes).

   answer: eosinophils

Platelets and Platelet Response

Platelets were formerly known as thrombocytes. The suffix "cytes" denotes that they are a cell. However, it has been found that they are not cells. They are actually little fragments of stem cells called **megakaryoblasts**. Megakaryoblasts are large cells that will fragment when they begin to leave the bone marrow and enter into circulation. These little fragments contain chemicals that will be released when it is time for blood clotting. Because these little fragments look like little plates (when they are highly magnified) they are called platelets. The formation of the platelets is still referred to as **thrombopoiesis,** however.

When a blood vessel is cut, it typically develops jagged edges. Platelets circulating in the blood will stick to those jagged edges. As the platelets stick they begin to pile on top of each other. The platelets on the bottom begin to burst open. When they do, they release a chemical called **platelet thromboplastin factor** (PTF). This chemical initiates the blood clotting process **(hemostasis).** There are a series of chemical reactions leading to the formation of a clot. These chemical reactions require several blood clotting factors. Table 14-2 discusses a few select blood clotting factors. The blood clotting factors are numbered with Roman numerals from I to XIII. However, blood clotting factor VI has been found to not be a part of the reactions. The factors are numbered according to when they were discovered. Anticoagulants (anticlotting agents) are chemicals that prevent blood from clotting by inhibiting the action of either the platelets or one of the blood clotting factors.

Table 14-2 Select Blood Clotting Factors	
Blood Clotting Factors	*Blood Clotting Factor Information*
Calcium ions (Ca^{2+})	This is blood clotting factor IV. **Heparin** is an anticlotting agent that inhibits factor IV. Calcium ions can be obtained from proper diet.
Fibrinogen	This is blood clotting factor I. This is one of the blood proteins mentioned earlier.
Antihemophiliac factor	This is blood clotting factor VIII. People lacking the gene for this factor have the disease known as **hemophilia.**
Christmas factor	This is blood clotting factor IX. This factor is produced by the liver. The liver needs vitamin K to make this factor. **Coumadin** is an anticlotting agent that prevents the liver from using vitamin K. This factor was named after a family with the last name of Christmas. This family lacked a blood clotting factor and researchers named it after them and it was the 9th factor discovered.

As we age (or because of what we eat), we begin to build up plaque in our arteries. This plaque has jagged edges. Therefore, platelets have a tendency to stick to those jagged edges. Once again, the bottom platelets burst and release platelet thromboplastin factor. This sets up the scenario for a blood clot. If this blood clot travels to the brain, a stroke may result. This is why many physicians suggest taking aspirin to reduce the chances of a cerebral clot. Aspirin acts as an anticlotting agent because it reduces the stickiness of the platelets. If the platelets can't stick to anything, they can't release platelet thromboplastin factor. Without this factor, the blood clotting process will not begin.

Example Problems

1. Thrombocytes are now known as _____.

    answer: platelets

2. Platelets are derived from bone marrow cells called _____.

    answer: megakaryoblasts

3. Coumadin does not change the consistency of blood and therefore really does not thin the blood. Rather, coumadin prevents blood from clotting by preventing the liver from using _____, which is necessary to make some of the blood clotting factors.

    answer: vitamin K

4. The condition of hemophilia is due to the lack of clotting factor number _____.

    answer: VIII

5. Which anticlotting agent reduces the stickiness platelets and therefore reduces blood clotting?

    answer: aspirin

Work Problems

1. The largest component of whole blood is _____.

2. Lymphocytosis is a term that means having too many lymphocytes. Lymphopenia is a term that means not having enough lymphocytes. Which term would apply if a patient had 15% lymphocytes?

3. Neutrophilia is a term that means having too many neutrophils. Neutropenia is a term that means not having enough neutrophils. Which term would apply if a patient had 82% neutrophils?

4. Aspirin is not a blood thinner because it does not change the consistency of blood. It prevents blood from clotting by doing what to the platelets?

5. True or false: The Christmas factor was identified on Christmas day.

6. A patient with an excess of eosinophils probably has either an allergy of some sort or an infestation of _____.

7. The process of red blood cell formation is called _____ and the process of white blood cell formation is called _____.

8. The _____ are the cells that fight infectious organisms.

9. Oxygen binds to hemoglobin to be transported to all tissues of the body. However, more specifically, oxygen binds to the _____ portion of the hemoglobin molecule.

10. The most common white blood cell circulating in a patient's blood who is in homeostasis is _____.

Worked Solutions

1. **plasma**

2. **lymphopenia**

3. **neutrophilia**

4. **Aspirin reduces the stickiness of the platelets.**

5. **false**

6. **worms**

7. **erythropoiesis; leukopoiesis**

8. **leukocytes**

9. **iron (Fe)**

10. **neutrophils**

Blood Types

In the 1800s, it was well understood that if a person lost blood, he or she might die. So, because it was thought that all blood was the same, physicians would take blood from anyone and use it for their patients. Most often, the patient died. It was then assumed that perhaps blood from one person was not necessarily the same as another person. By 1933, it was found that different people's blood had different glycolipids on their blood cell membrane. The first cell studied had a specific glycolipid and was given the identification letter A (type A blood). The second cell studied had a specific glycolipid and was given the identification letter B (type B blood). The third cell studied had a glycolipid that was identical to those with type A and also a glycolipid that was identical to those with type B. These erythrocytes were identified as AB. The fourth cell studied was lacking those specific glycolipids and was given the identification as O.

Glycolipids and Blood Typing

The glycolipids on the surface of cell membranes are known as antigens. A more precise term when discussing blood cells is **agglutinogen.** In order to understand who can donate to whom, we must have an understanding of the glycolipids on the surface of the cell membrane. We must also have an understanding of the types of protein molecules found in the plasma of the blood. The plasma proteins (that are associated with blood typing) are called antibodies or more precise; **agglutinins.** Table 14-3 discusses pertinent information regarding agglutinogens and agglutinins.

Table 14-3 Blood Typing Information								
Whole Blood			**Packed Cells**			**Plasma**		
Consists of erythrocytes and plasma			Consists of primarily erythrocytes.			Consists of primarily plasma.		
The erythrocytes have the agglutinogens. The plasma has the agglutinins.			Consists of agglutinogens.			Consists of agglutinins.		
Blood type	**Agglu-tinogen**	**Agglu-tinin**	**Blood type**	**Agglu-tinogen**	**Agglu-tinin**	**Blood type**	**Agglu-tinogen**	**Agglu-tinin**
A	A	b	A	A	–	A	–	b
B	B	a	B	B	–	B	–	a
AB	A and B	none	AB	A and B	none	AB	–	None
O	none	a and b	O	none	–	O	none	a and b

Donating Packed Cells

Whenever an agglutinin binds to an agglutinogen we say it is activated and this binding action will cause the erythrocytes to clump and could cause death. Therefore, at all costs, we need to prevent the agglutinins from binding to the agglutinogens. Figure 14-2 shows the donation of packed cells to a patient. When donating packed cells, only the erythrocyte with its agglutinogen is donated.

In this scenario, a person with type B blood is going to donate to a patient with type A blood. In this case, we are going to donate packed cells. Therefore, the only thing that will be donated is the erythrocyte (agglutinogen), not the agglutinin. So, put an X over the "a agglutinin," because it is not going to be donated. Look at Figure 14-2b.

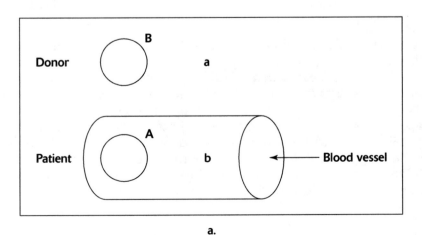

a.

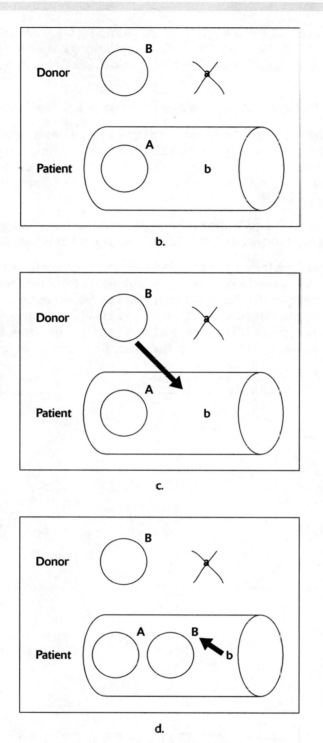

Figure 14-2: Donation of packed cells

Now, the only thing to be donated is the erythrocyte with the B agglutinogen. Draw an arrow from the donor's erythrocyte to the inside of the blood vessel as shown in Figure 14-2c.

Draw the donated erythrocyte inside the blood vessel. Then, check to see if there are any agglutinogens that have the same "letter" as the agglutinin. In this case, the agglutinogen is the same letter as the agglutinin. Both "letter Bs" are in the same blood vessel. The b agglutinin will bind to the B agglutinogen and will cause blood cells to clump. We typically say that the b agglutinin has been activated. To show the binding, draw an arrow from the b agglutinin to the B agglutinogen. Examine Figure 14-2d.

When the b agglutinin binds to the B agglutinogen, clumping **(agglutination)** will occur. When blood cells clump, they will burst **(hemolyze).** If the agglutinogens inside the blood vessel have a different "letter" than the agglutinins in the blood vessel, clumping will not occur and the donation will be safe.

Examine Figure 14-3 a through d to see another example of a packed cell donation.

In this scenario, a person with type AB blood is going to donate to a patient with type B blood. In this case, we are going to donate packed cells. Therefore, the only thing that will be donated is the erythrocyte (agglutinogen), not the agglutinin. So, put an X over the "agglutinin," because it is not going to be donated. However, in this case, the donor's plasma doesn't contain any agglutinins. Look at Figure 14-3b.

Now, the only thing to be donated is the erythrocyte with the A and B agglutinogen. Draw an arrow from the donor's erythrocyte to the inside of the blood vessel as shown in Figure 14-3c.

Draw the donated erythrocyte inside the blood vessel. Then, check to see whether there are any agglutinogens that have the same "letter" as the agglutinin. In this case, one of the agglutinogens is the letter as the agglutinin. Both "letter As" are in the same blood vessel. The a agglutinin will bind to the A agglutinogen and will cause blood cells to clump. We typically say that the a agglutinin has been activated. To show the binding, draw an arrow from the a agglutinin to the A agglutinogen. Examine Figure 14-3d. In this case, the donated erythrocytes will agglutinate and hemolyze.

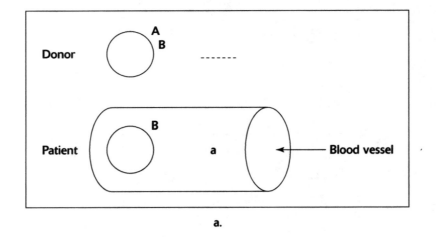

a.

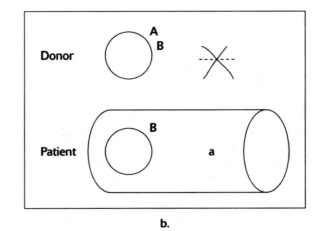

b.

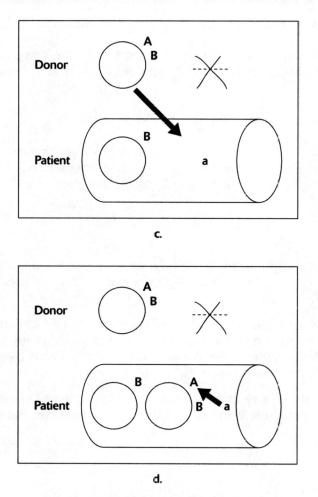

c.

d.

Figure 14-3: Donation of packed cells.

Example Problems

Follow the steps identified in Figure 14-2 and 14-3 to answer the following questions.

1. Can type B packed cells safely donate to a type AB patient?

    answer: yes; In this case, you are donating the B agglutinogen. The AB patient does not have any agglutinins so there isn't anything to bind to the B agglutinogen.

2. Can type B packed cells safely donate to a type O patient?

    answer: no; In this case, you are donating the B agglutinogen. The type O patient has the a and b agglutinins. The b agglutinins will bind to the B agglutinogens. Clumping will occur.

3. Can type O packed cells safely donate to a type A patient?

    answer: yes; In this case, you are donating the erythrocyte that doesn't have any agglutinogens. Because there aren't any agglutinogens, there isn't anything for the b agglutinins (from the A patient) to bind to.

4. Can type AB packed cells safely donate to a type O patient?

   answer: no; In this case, you are donating A and B agglutinogens. The type O patient has the a and b agglutinins. Both the a and b agglutinins will bind to the A and B agglutinogens. Clumping will occur.

5. Can type O packed cells safely donate to a type AB patient?

   answer: yes; In this case, you are donating an erythrocyte that doesn't have any agglutinogens. Also in this case, the type AB patient doesn't have any agglutinins. There isn't anything to bind to any agglutinogens and there aren't any agglutinogens to bind to anyway.

Donating Whole Blood

Examine Figures 14-4a through 14-4d to see how whole blood is donated. When donating whole blood, both the erythrocyte (with its agglutinogen) and the plasma (with its agglutinins) are part of the donation.

In this scenario, a person with type B blood is going to donate to a patient with type A blood. In this case, we are going to donate whole blood. Therefore, we are going to donate not only the agglutinogen but also the agglutinin. So, we do not need to put an X over the "a agglutinin," because it is going to be donated. Look at Figure 14-4b.

Draw an arrow from the donor's erythrocyte to the inside of the blood vessel as shown in Figure 14-4c. Also, draw an arrow from the donor's a agglutinin into the patient's blood vessel.

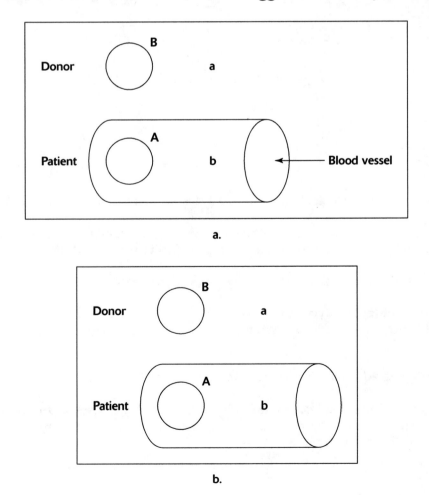

a.

b.

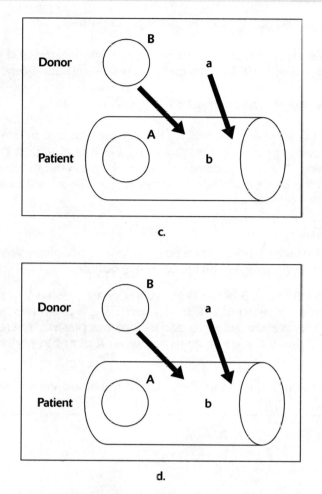

Figure 14-4: Donation of whole blood.

Draw the donated erythrocyte inside the blood vessel. Then, write the letter a in the blood vessel to represent the donation of the agglutinin. Then, check to see if there are any agglutinogens that have the same "letter" as the agglutinin. In this case, the agglutinogen is the same letter as the agglutinin. Both "letter Bs" are in the same blood vessel. The b agglutinin will bind to the B agglutinogen and will cause blood cells to clump. We typically say that the b agglutinin has been activated. Also, there is a capital A and a lower case a in the same blood vessel. To show the binding, draw an arrow from the b agglutinin to the B agglutinogen and from the a agglutinin to the A agglutinogen. Examine Figure 14-4d.

When the b agglutinin binds to the B agglutinogen, clumping (agglutination) will occur. When the a agglutinin binds to the A agglutinogen, clumping will occur. When blood cells clump, they will burst (hemolyze). If the agglutinogens inside the blood vessel have a different "letter" than the agglutinins in the blood vessel, clumping will not occur and the donation will be safe.

Example Problems

Follow the steps identified in Figure 14-4 to answer the following questions.

1. Can type B whole blood safely donate to a type AB patient?

   answer: no; In this case, you are donating the B agglutinogen and the a agglutinin. The donated a agglutinin will bind to the A agglutinogen of the patient.

2. Can type A whole blood safely donate to a type O patient?

    answer: no; In this case, you are donating the A agglutinogen and the b agglutinin. The a agglutinin of the patient will bind to the donated A agglutinogen. Blood will clump.

3. Can type O whole blood safely donate to a type A patient?

    answer: no; In this case, you are donating the erythrocyte that doesn't have any agglutinogens. However, you are donating the a and b agglutinins. The type A patient has the b agglutinins. The b agglutinins will not bind to anything. However, the donated a agglutinins will bind to the patient's A agglutinogen. Blood will clump.

Donating Plasma

Examine Figures 14-5a through 14-5d to see how plasma is donated. When donating plasma, only the plasma (with its agglutinins) is included in the donation.

In this scenario, a person with type B blood is going to donate plasma to a patient with type A blood. This type of donation is worded in this manner: Type B plasma is being donated to a type A patient. In this case, we are going to donate just the plasma. Therefore, we are going to donate only the agglutinin. So, we will need to put an X over the erythrocyte of the donor. Look at Figure 14-5b.

Draw an arrow from the donor's plasma to the inside of the blood vessel as shown in Figure 14-5c.

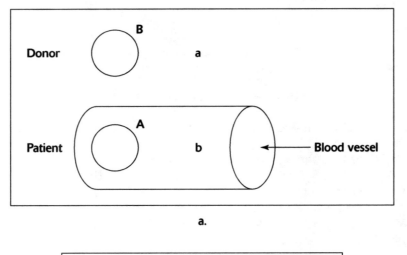

a.

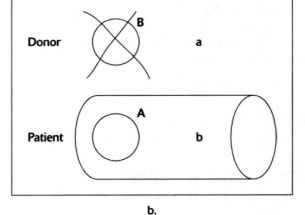

b.

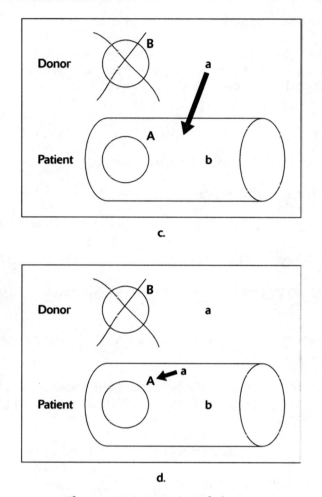

Figure 14-5: Donation of plasma.

Write the letter a in the blood vessel to represent the donation of the agglutinin. Then, check to see if there are any agglutinogens that have the same "letter" as the agglutinin. In this case, the agglutinogen (of the patient) is the same letter as the donated agglutinin. To show the binding, draw an arrow from the a agglutinin to the A agglutinogen. Examine Figure 14-5d.

When the a agglutinin binds to the A agglutinogen, clumping (agglutination) will occur. When blood cells clump, they will burst (hemolyze). If the agglutinogens inside the blood vessel have a different "letter" than the agglutinins in the blood vessel, clumping will not occur and the donation will be safe.

Example Problems

Follow the steps identified in Figure 14-5 to answer the following questions.

1. Can type A plasma be donated to a type B patient?

    answer: no; In this case, you are donating the b agglutinin. The donated b agglutinin will bind to the B agglutinogen of the patient Blood will clump.

2. Can type AB plasma be donated to a type A patient?

   answer: yes; In this case, you are donating the plasma of a type AB person. AB people do not have any agglutinins. The donated plasma does not have any agglutinins to bind to the agglutinogens of the patient. This is a safe donation.

Work Problems

1. Can type AB whole blood donate to a type O patient?

2. Can type O packed cells donate to a type B patient?

3. Can type A plasma be donated to a type AB patient?

4. Why is a person with type O packed cells considered to be the universal donor?

5. Type AB people can only receive blood from other AB people because they have both kinds of _____.

6. When we are donating type A plasma, we are actually donating the _____ of a type A donor.

7. Type A plasma cannot be donated to a type B patient because the plasma from the donor contains the _____ agglutinins, which will bind to the agglutinogens of the patient.

8. When whole blood is being donated, both the _____ and _____ are being donated to the patient.

9. In the case of packed cell donations, the _____ are not being donated.

10. In the case of plasma donations, the _____ are not being donated.

Worked Solutions

1. **no.** The a and b agglutinins of the type O patient will bind to the A and B agglutinogens of the donor.

2. **yes.** The a agglutinins of the patient do not have anything to bind to from the donor. The donor doesn't have any agglutinogens.

3. **no.** The b agglutinins of the donor will bind to the B agglutinogens of the patient.

4. **Type O packed cells is the universal donor because a type O person does not have any agglutinogens.** Without agglutinogens, there isn't anything for agglutinins, in the patients, to bind to.

5. **agglutinogens**

6. **agglutinins**

7. **b**

8. **agglutinogen; agglutinin**

9. **agglutinins**

10. **agglutinogens**

The Rh factor

There are numerous agglutinogens on the surface of erythrocytes. By far the most common are classified as the ABO group. Each time an agglutinogen is discovered, it is given a letter of the alphabet, typically in the sequence in which they were discovered. As time went on, there was a C agglutinogen and then a D agglutinogen, etc. The D agglutinogen has been dubbed as the **Rh factor.** This is because the D agglutinogen was first discovered in a group of monkeys with the genus name Rhesus. People who have the D agglutinogen on their erythrocyte are said to have positive blood. People who don't have the D agglutinogen are said to have negative blood. The D agglutinogen has gained a lot of prominence due to its effects during pregnancy. Figure 14-6 illustrates the action of the D agglutinogen.

In this scenario, we are donating type B+ packed cells to a person with type B– blood. Draw an arrow from the donor's erythrocyte to the inside of the blood vessel. Look at Figure 14-6b.

Draw the B+ erythrocyte inside the blood vessel. Then look to see if there are any agglutinins that will bind to the agglutinogens. Look at Figure 14-6c.

This is a safe donation. B+ packed cells can be donated to a B– person. However, over a period of time, the B– person will begin to manufacture the d agglutinins. The B+ donated blood will die and decompose in about 120 days. Figure 14-6d shows the manufactured d agglutinins.

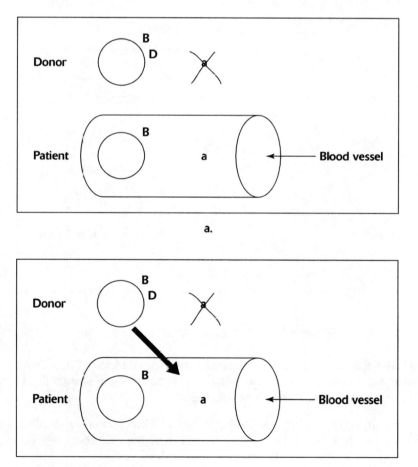

a.

b.

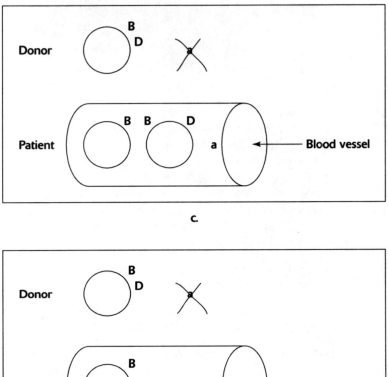

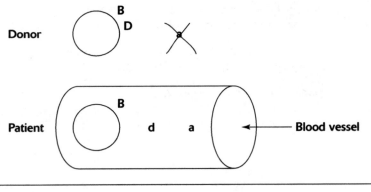

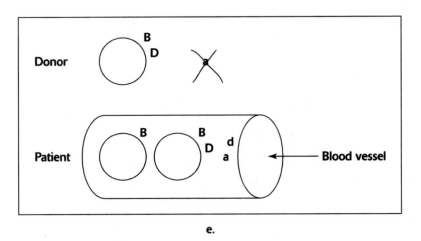

Figure 14-6: Donations involving the Rh factor.

Now, some time later, the same B– patient needs some more blood. Let's donate some B+ blood to this person. After all, it was a safe donation before. Figure 14-6e shows the donated B+ blood in the patient's blood vessel after the second donation.

Now, examine to see if there are any agglutinins that will bind to an agglutinogen. In this case, the d agglutinin that was manufactured from the first positive donation will bind to the D agglutinogen from the donor's blood. This is now an unsafe donation.

Many years ago, it was said that a negative patient could receive positive blood only once. Since 1970, however, a negative patient can receive positive blood several times. This is because of the development of a drug called **RhoGam.**

When a negative patient receives positive blood, they are given a shot of RhoGam. RhoGam does not prevent the negative patient from manufacturing the d agglutinin but rather it will mask the d agglutinins when they are produced. Therefore, because the d agglutinins are masked, they will not bind to anything and therefore act as though they don't exist.

The Rh factor was a grave concern to pregnant women prior to 1970. If the expectant mother has negative blood and the developing baby has positive blood, the positive blood would (at the time of birth) enter into the mother's bloodstream. This would be just like a positive donation given to a negative patient. The expectant mother will begin to manufacture the d agglutinin. Now, mother has the d agglutinin in her plasma. If any of the d agglutinins should happen to enter into the baby's circulatory system, there will be a combination of the D agglutinogen and d agglutinin in the baby's bloodstream. Agglutination will occur in the baby and cause death to the child. If the baby dies before birth, it is known as **erythroblastis fetalis** and if the baby dies after birth, it is known as **hemolytic disease of the newborn** (HDN).

But, today there is not a problem. If the expectant mother has negative blood, she is given a shot of RhoGam during pregnancy. RhoGam will mask any d agglutinins the mother may produce before birth. She is also given a second shot of RhoGam right after the birth of the child to mask any d agglutinins that might be made at that point in time due to "leakage" of the baby's blood through the placenta into the mother's circulatory system.

Example Problems

The following questions are in reference to the Rh factor.

1. Can an A– patient receive A+ blood a second time prior to 1970?

    answer: no; The A– patient would have manufactured the d agglutinins from the first positive donation. Those d agglutinins will bind to the D agglutinogens from the donor's blood from the second donation.

2. Can an AB– patient receive AB+ blood a second time after 1970?

    answer: yes; RhoGam was developed and widely used after 1970. RhoGam would mask any d agglutinins produced from the first donation of positive blood and therefore would ensure that the second donation would be safe.

3. RhoGam got its name from _____.

    answer: research conducted on Rhesus monkeys

Work Problems

1. Can type AB+ packed cells donate to a type AB– patient a second time since the advent of RhoGam?

2. True or false: RhoGam prevents an Rh– patient from producing the d agglutinins.

3. Prior to RhoGam, a problem may have existed during pregnancy if the expectant mother had Rh _____ blood and the first born had Rh _____ blood.

4. Who has the ability to manufacture the d agglutinins (Rh+ people or Rh– people)?

5. Prior to RhoGam, an Rh– person would develop the d agglutinins only if they were exposed to what kind of blood?

6. Hemolytic disease of the newborn is due to the D agglutinogens of the baby being exposed to the _____ from mother's blood.

7. In the Caucasian population, approximately 70% of the population has positive blood. This means they have the _____ agglutinogen.

8. A person who is considered to have A+ blood has the _____ agglutinogen and the _____ agglutinogen.

9. A person who is considered to have B– blood is lacking the _____ agglutinogen.

10. What kind of agglutinogens does an O+ person have?

Worked Solutions

1. **yes.** Because the development of RhoGam, any agglutinins produced by the AB– patient will be masked.

2. **false.** RhoGam will only mask the produced d agglutinins.

3. **negative; positive**

4. **Rh– people are the only ones who have the genetics to manufacture the d agglutinins.**

5. **Exposure to Rh+ blood would cause the Rh– people to manufacture the d agglutinins.**

6. **d agglutinins**

7. **D agglutinogen**

8. **A; D**

9. **D**

10. **An O+ person has only the D agglutinogen.**

Chapter Problems and Solutions

Problems

1. Erythropoietin is produced by the _____.

2. The stem cells of red blood cells are called _____.

3. Plasma makes up about _____% of whole blood.

4. The major component of plasma is _____.

5. A cubic mm of blood contains about _____ erythrocytes.

6. What triggers the release of erythropoietin? _____

7. Type A packed cells cannot donate to a type B patient because the _____ of the B patient will bind to the _____ of the donor.

8. Type O packed cells is considered to be the universal donor because the type O donor lacks _____.

9. Type AB would be considered the universal donor of plasma because their plasma does not contain any _____.

10. Only Rh _____ people have the ability to manufacture the d agglutinins.

Answers and Solutions

1. **kidneys**

2. **erythroblasts**

3. **55%**

4. **water (92%)**

5. **5 million**

6. **a lack of oxygen going to the kidney cells**

7. **a agglutinins; A agglutinogens**

8. **agglutinogens**

9. **agglutinins**

10. **negative.** If Rh+ people had the genetics to manufacture the d agglutinin, they would have agglutination within their own blood.

Supplemental Chapter Problems

Problems

1. An excess number of leukocytes are called leukocytosis. Therefore, an excess number of monocytes would be called _____.

2. A WBC count that shows a reduced number of leukocytes is called leukopenia. Therefore, a low number of lymphocytes would be called _____.

3. A laboratory technologist did a differential count of the leukocytes and determined there to be 15 lymphocytes out of 100 cells counted. Is this value too high or too low regarding normal lymphocytes counts?

4. When basophils release _____ during an allergy response, they typically cause the blood vessels lining the nasal cavity to dilate thus resulting in a runny nose.

5. When oxygen binds to the _____ portion of hemoglobin, it causes the blood to become a bright red color.

6. Carbon monoxide is a deadly gas. It can cause death because carbon monoxide has an affect on the respiratory centers and also causes the blood cells to basically deliver less oxygen to the tissues of the body. This is because carbon monoxide binds to the _____ portion of the hemoglobin just as oxygen does.

7. The process of red blood cell formation is called _____.

8. The process of blood clotting is called _____.

9. If the body experiences a condition known as hypoxia, this will cause the kidneys to begin releasing _____.

10. Excessive aspirin intake could cause failure of the _____ mechanism.

Answers

1. monocytosis

2. lymphopenia

3. This turns out to be 15% lymphocytes. A normal value is 20–40%. A value of 15% would be too low. This is known as lymphopenia.

4. histamine

5. iron

6. iron

7. erythropoiesis

8. hemostasis

9. erythropoietin

10. blood clotting

Chapter 15
The Heart and Blood Vessels

Chapter 14 discusses the numerous functions of the blood, such as transporting oxygen and nutrients to tissues and transporting hormones to various tissues, so that the body can respond in a specified manner. This chapter discusses how the heart is involved in pumping the blood to all the tissues, so that the blood can carry out its function. As the heart pumps, it forces the blood into a whole array of blood vessels.

Structure of the Internal Heart

The best way to study the structure of the heart is to take a drop of blood and follow it through the heart. As the blood is traveling through the heart, identify the structures it passes through. Figure 15-1 shows the internal structures of the heart and each part is identified with a number; refer to it as you read the step-by-step flow of blood through the heart.

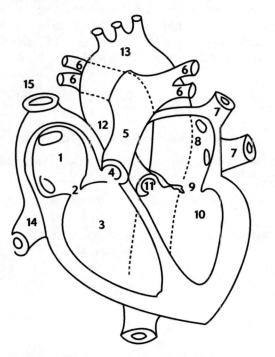

Figure 15-1: Internal structures of the heart.

1. Begin with blood in chamber number 1, the **right atrium.**

2. The blood will then pass through structure number 2, the **tricuspid valve.**

3. The blood passes through the valve and enters into chamber number 3, the **right ventricle.**

4. From there, the blood is pumped upward toward valve number 4, the **pulmonary semilunar valve.**

5. The blood passes through the valve and enters number 5, the **pulmonary trunk.**

6. From the pulmonary trunk, the blood enters number 6, the **pulmonary arteries.**

7. The blood in the pulmonary arteries is on its way to the lungs to pick up oxygen.

8. After the blood picks up oxygen in the lungs, it returns to the heart by traveling through number 7, the **pulmonary veins.**

9. The pulmonary veins will transport the blood to chamber number 8, the **left atrium.**

10. The blood will then pass through structure number 9, the **bicuspid valve.**

11. The blood passes through the valve and enters into chamber number 10, the **left ventricle.**

12. From there, the blood is pumped upward toward valve number 11, the **aortic semilunar valve.**

13. The blood passes through the valve and enters structure number 12, the **ascending aorta.**

14. From there, the blood continues on into number 13, the **aortic arch.**

15. Blood in the aortic arch will enter into several vessels. In essence, at this point, the blood in the aortic arch is on its way to the head, arms, chest, and lower body. As the blood reaches the rest of the body, it delivers oxygen to the tissues and picks up the waste product, carbon dioxide (CO_2) that is produced by the tissues. The blood now returns to the heart via two major vessels.

16. From the lower body, the blood returns to the heart via vessel number 14, the **inferior vena cava.**

17. Blood in the inferior vena cava (IVC) will enter the right atrium.

18. From the upper body, the blood returns to the heart via vessel number 15, the **superior vena cava.**

19. Blood in the superior vena cava (SVC) will enter the right atrium.

Figure 15-2 is a flow chart showing the blood flow from ascending aorta and the aortic arch. Figure 15-3 identifies the vessels mentioned in the flow chart.

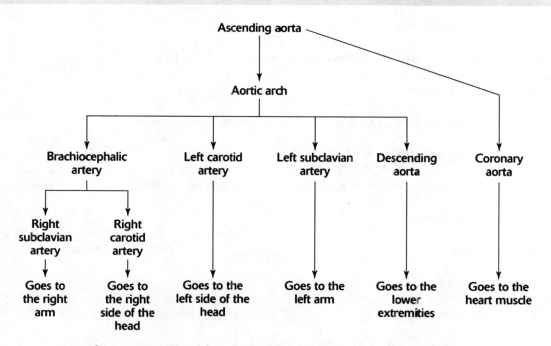

Figure 15-2: Blood flow from the ascending aorta and aortic arch.

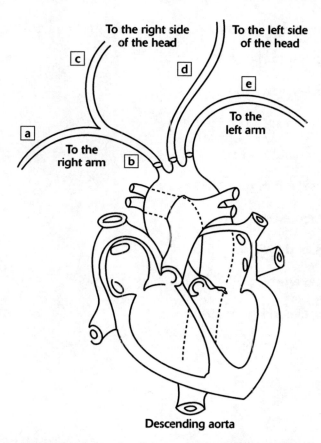

Figure 15-3: Blood vessels of the ascending aorta and aortic arch.

Example Problems

1. Blood in the aortic arch came from which chamber of the heart?

    answer: left ventricle

2. Blood entering the lungs came from which chamber of the heart?

    answer: right ventricle

3. Blood passing through the tricuspid valve came from which chamber of the heart?

    answer: right atrium

4. Blood in the pulmonary _____ is on its way to the lungs.

    answer: arteries

5. Blood in the pulmonary _____ is on its way back to the heart from the lungs.

    answer: veins

Answer the following questions based on the information in **Figure 15-2** and **Figure 15-3**.

1. Identify blood vessel a in **Figure 15-3**.

    answer: right subclavian artery

2. Identify blood vessel b in **Figure 15-3**.

    answer: brachiocephalic artery

3. Identify blood vessel c in **Figure 15-3**.

    answer: right carotid artery

4. Identify blood vessel d in **Figure 15-3**.

    answer: left carotid artery

5. Identify blood vessel e in **Figure 15-3**.

    answer: left subclavian artery

Other Features of the Internal Heart

The atria of the heart contracts and forces the blood through the **atrioventricular valves** (tricuspid and bicuspid valves). Due to the position and shape of the valves, blood naturally flows into the two ventricles. The valves form a one-way door into the ventricles. Once in the ventricles, the blood is pumped upward from the **apex.** As blood is forced upward, it pushes on the atrioventricular valves thus forcing them closed. Now, the only place for the blood to go is in the two semilunar valves. The semilunar valves are also shaped to form a one-way door into the ascending aorta or the pulmonary trunk.

Because blood is forced against the atrio-ventricular valves, there needs to be some mechanism to prevent the valves from inverting back into the atria. Notice that those valves have chords attached to them. These are the **chordae tendineae** (letter a in Figure 15-4). These chords are attached to special muscles in the ventricles called **papillary muscles** (letter b in Figure 15-4). The papillary muscles contract in such a manner to prevent the valves from inverting. If the valves don't close properly, blood may pass back into the atria (going backwards through the valves). As the blood passes in reverse through the valves, it creates a sound. This is termed as a **heart murmur.**

Located only in the right ventricle is the **moderator band** (letter c in Figure 15-4). The moderator band serves to "moderate" the size of the right ventricle. Notice that the right ventricle has a thinner wall than the left ventricle. Because the right ventricle has thin walls, it has a tendency to expand too much when blood from the right atrium enters the ventricle. The moderator band serves to prevent "hyperexpansion" of the right ventricle.

The left ventricle has a much thicker wall than the right ventricle. Blood in the left ventricle has to be pumped out to all parts of the body. Therefore, a strong muscular pump is required. Blood in the right ventricle only has to travel to the lungs. The lungs are in close proximity to the heart. Therefore, a weaker pump is sufficient to get the blood to the lungs. Examine Figure 15-4.

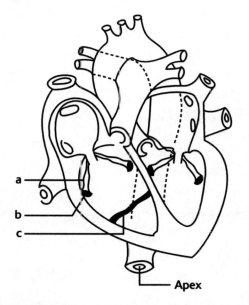

Figure 15-4: Other Internal structures of the heart.

The Conducting System of the Heart

There are special cells of the heart that are responsible for getting the heart to contract. Embedded in the right atrium near the entrance of the superior vena cava is the **sinoatrial node** (SA node). The sinoatrial node generates impulses that spread across the atria. As the impulse spreads across the atria, it causes the atria to contract thereby forcing blood toward the ventricles. The impulse arrives at the **atrioventricular node** (AV node). This group of cells is located at the junction of the atria and the ventricles. From the AV node, the impulse travels down the ventricular septum of the heart via the **bundle branches.** Once the impulse reaches the apex region of the heart, the impulse spreads through the ventricles via the **Purkinje fibers.** Once the Purkinje fibers have been activated, the ventricles will contract from the apex upward. This action forces the blood from the ventricles into the semilunar valves.

Figure 15-5 shows the heart and the conducting system of the heart, with the various parts of the conducting system identified with a letter. Answer the following questions based on the letters on Figure 15-5.

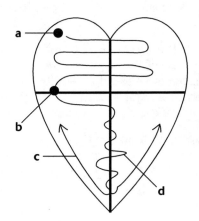

Figure 15-5: The conducting system of the heart.

Example Problems

Answer the following questions by examining Figure 15-5.

1. Which letter represents the sinoatrial node?

   answer: a

2. Which letter represents the atrioventricular node?

   answer: b

3. Which letter represents the bundle branches?

   answer: d

4. Which letter represents the Purkinje fibers?

   answer: c

The Electrocardiogram (ECG)

A recording of the electrical activities of the conduction system of the heart constitutes an **electrocardiogram**. An electrocardiogram consists of numerous bumps and valleys. Each of those bumps and valleys illustrates the activity of the heart and relays significant information to a cardiologist. Figure 15-6 shows a sample ECG. The following bullet list gives you select information regarding those bumps and valleys.

- ❏ **P wave:** Represents atrial depolarization, which correlates with atrial contraction. It also represents the impulse activity from the SA node to the AV node.

- ❏ **QRS wave:** Represents ventricular depolarization, which correlates with ventricular contraction. It also represents the impulse along the Purkinje fibers; atrial repolarization occurs during this time.

- ❏ **T wave:** Represents ventricular repolarization.

- ❏ **P-Q segment:** Represents the impulse traveling down the bundle branches.

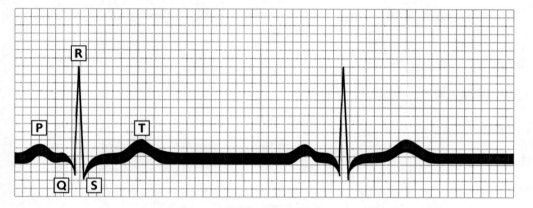

Figure 15-6: A sample ECG recording.

Work Problems

1. The heart consists of (how many) _____ chambers.

2. The two superior chambers of the heart are called _____ and the two inferior chambers are called _____.

3. The tricuspid valve is located on the _____ side of the heart.

4. What is the name of the two atrioventricular valves?

5. Which vessels are transporting oxygenated blood back to the heart?

6. Which node of the heart is considered to be the pacemaker?

7. When activated, which nerve fibers cause the ventricles to contract?

8.   The QRS complex on an ECG represents the depolarization of the _____.

9.   The moderator band is located only in the _____.

10. What causes the bicuspid and tricuspid valve to close?

Worked Solutions

1.   **4**

2.   **atria; ventricles**

3.   **right**

4.   **tricuspid; bicuspid (mitral)**

5.   **pulmonary veins.** The IVC and SVC are transporting deoxygenated blood back to the heart.

6.   **sinoatrial node (SA node)**

7.   **Purkinje fibers**

8.   **ventricles**

9.   **right ventricle**

10. **The force of blood due to the contraction of the ventricles will cause the valves to close as the blood is forced from the apex up.**

More Heart Information

Here is additional information about the layers of the heart and its position in the body.

❑ The heart walls are made of three layers:

The **epicardium** is the outermost layer.

The **endocardium** is the innermost layer.

The **myocardium** is the middle layer and consists of cardiac cells.

❑ The base of the heart is the most superior portion near the aortic arch and pulmonary trunk.

❑ The apex of the heart is the most inferior portion and is the "pointy" part of the heart.

❑ The apex of the heart angles to the left of the midline of the body.

❑ The aortic arch arches to the left of the heart and becomes the thoracic aorta on the posterior side of the heart.

❑ The pulmonary trunk angles to the left of the heart.

Blood Vessels (Arteries)

Arteries are blood vessels that transport blood away from the heart. Most arteries transport oxygenated blood but there are some arteries that carry deoxygenated blood. The pulmonary arteries carry deoxygenated blood away from the heart and toward the lungs.

The best way to study the blood vessels is to take a drop of blood and follow it as it flows through the various arteries and veins of the body. Figure 15-7 shows the flow of blood exiting the major vessels associated with the heart.

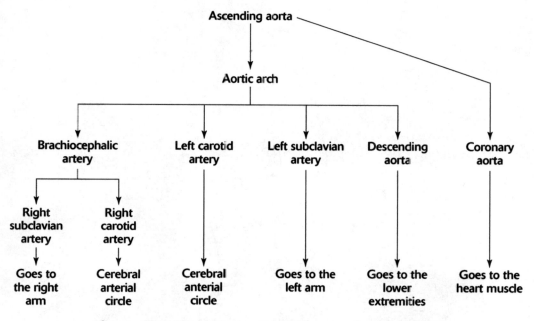

Figure 15-7: Chart showing blood flow away from the heart.

The blood in the right carotid artery enters the **cerebral arterial circle** on the right side. The cerebral arterial circle supplies blood to the pituitary gland. The left carotid artery enters the cerebral arterial circle on the left side. This organization ensures that the pituitary gland always receives a rich supply of blood. Figure 15-8 shows the blood flow away from the heart.

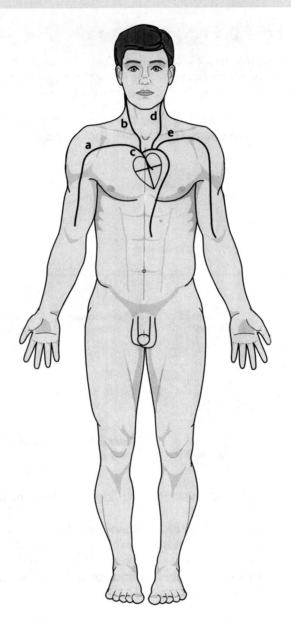

Figure 15-8: Blood flow away from the heart.

Example Problems

Answer the following questions by examining Figure 15-7 and Figure 15-8.

1. Which letter represents the brachiocephalic artery?

   answer: c

2. Which letter represents the right carotid artery?

   answer: b

3. Which letter represents the right subclavian artery?

   answer: a

4. Which letter represents the left subclavian artery?

   answer: e

5. Which letter represents the left carotid artery?

   answer: d

Figure 15-9 shows the flow of blood from the heart to the arms. Figure 15-10 is a picture that illustrates Figure 15-9. The brachial artery branches in the antecubital region to form the ulnar and radial artery.

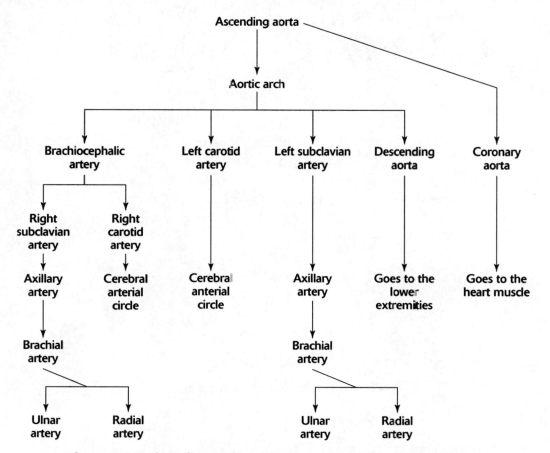

Figure 15-9: Chart showing blood flow away from the heart to the arms.

Figure 15-10 shows the flow of blood from the heart to the arms.

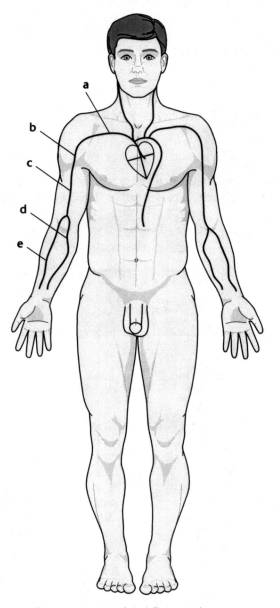

Figure 15-10: Blood flow to the arms.

Example Problems
Answer the following questions by examining Figure 15-9 and Figure 15-10.

1. Which letter represents the subclavian artery?

   answer: a

2. Which letter represents the axillary artery?

   answer: b

3. Which letter represents the brachial artery?

   answer: c

4. Which letter represents the radial artery?

   answer: e

5. Which letter represents the ulnar artery?

   answer: d

Figures 15-11 and 15-12 show the flow of blood from the heart to the legs.

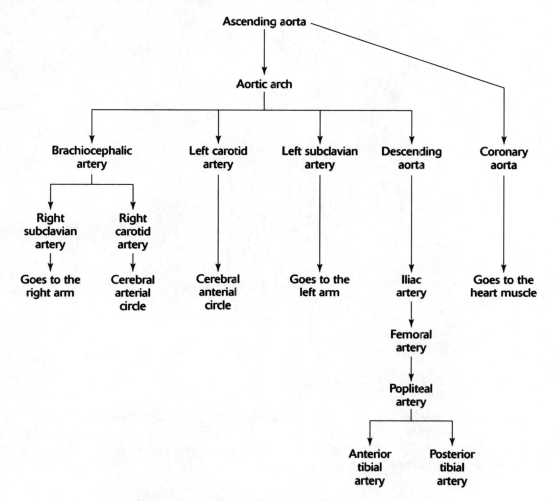

Figure 15-11: Chart showing blood flow to the legs.

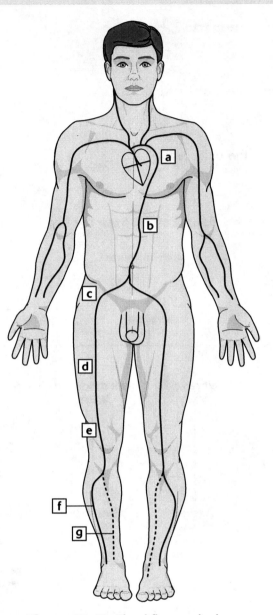

Figure 15-12: Blood flow to the legs.

The descending aorta can be divided into the thoracic aorta and abdominal aorta. The flow of blood from the aortic arch is as follows: aortic arch to the thoracic aorta to the abdominal aorta and then to the iliac artery. The popliteal artery branches in the popliteal region of the leg and forms the anterior and posterior tibial artery.

Example Problems

Answer the following questions by examining Figure 15-11 and Figure 15-12.

1. Which letter represents the thoracic aorta?

   answer: a

2. Which letter represents the abdominal aorta?

   answer: b

3. Which letter represents the iliac artery?

   answer: c

4. Which letter represents the femoral artery?

   answer: d

5. Which letter represents the popliteal artery?

   answer: e

6. Which letter represents the anterior tibial artery?

   answer: f

7. Which letter represents the posterior tibial artery?

   answer: g

Blood Vessels (Veins)

Veins are blood vessels that transport blood back to the heart. Most veins carry deoxygenated blood but the pulmonary veins are returning to the heart from the lungs and are therefore carrying oxygen. Figure 15-13 shows the flow of blood from the head region back to the right atrium of the heart. Figure 15-14 illustrates that same information. The right and left brachiocephalic veins join together to form the superior vena cava.

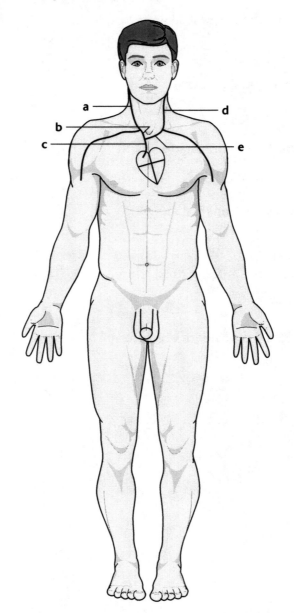

Figure 15-13: Chart showing blood flow from the head to the heart.

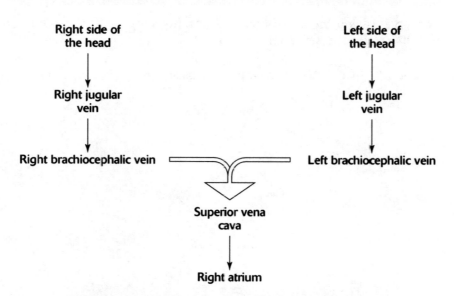

Figure 15-14: Blood flow from the head to the heart.

Example Problems

Answer the following questions by examining Figure 15-13 and Figure 15-14.

1. Which letter represents the superior vena cava? _____

   answer: c

2. Which letter represents the right brachiocephalic vein? _____

   answer: b

3. Which letter represents the left brachiocephalic vein? _____

   answer: e

4. Which letter represents the right jugular vein? _____

   answer: a

5. Which letter represents the left jugular vein? _____

   answer: d

Figure 15-15 is a chart showing the flow of blood from the arms back to the right atrium of the heart. Figure 15-16 illustrates the same information.

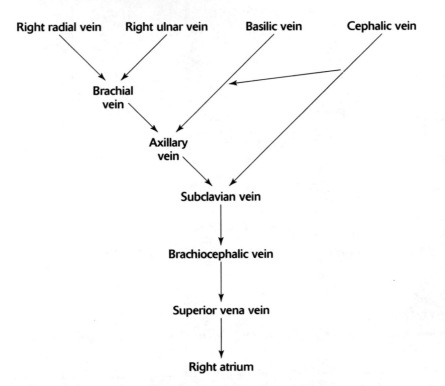

Figure 15-15: Chart showing blood flow from the arms to the heart.

The radial and ulnar veins join to form the brachial vein. The basilic vein runs the length of the arm to the axillary vein. The cephalic vein also runs the length of the arm but joins with the subclavian vein. Blood in the cephalic vein will also enter into the basilic vein. The vein that crosses over to the basilic about the region of the antecubital is called the cubital vein. It's called the cubital vein even though it's in the antecubital region.

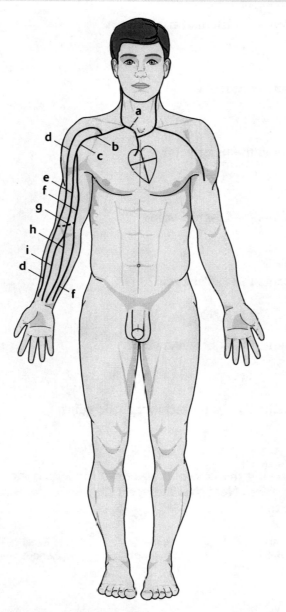

Figure 15-16: Blood flow from the arms to the heart.

Example Problems

Answer the following questions by examining Figure 15-15 and Figure 15-16.

1. Which letter represents the cephalic vein?

    answer: d

2. Which letter represents the basilic vein?

    answer: f

3. Which letter represents the median cubital vein?

    answer: g

4. Which letter represents the radial vein?

    answer: i

5. Which letter represents the ulnar vein?

    answer: h

6. Which letter represents the brachial vein?

    answer: e

7. Which letter represents the axillary vein?

    answer: c

8. Which letter represents the subclavian vein?

    answer: b

9. Which letter represents the right brachiocephalic vein?

    answer: a

Figure 15-17 is a chart showing the flow of blood from the legs back to the heart. Figure 15-18 illustrates the same information. Note that the great saphenous vein runs from the lower leg all the way to the femoral vein.

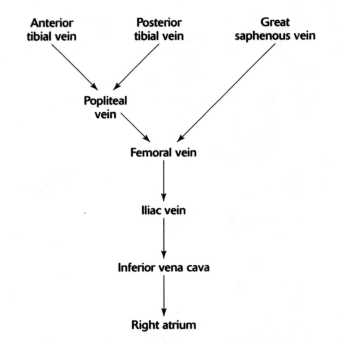

Figure 15-17: Chart showing blood flow from the legs to the heart.

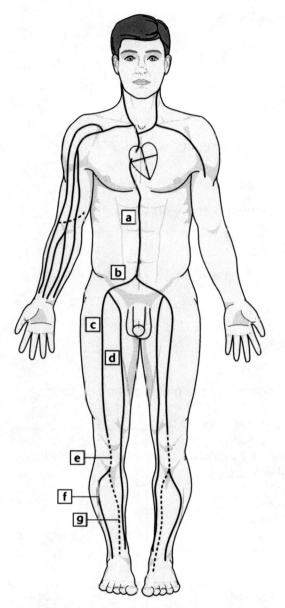

Figure 15-18: Blood flow from the legs to the heart.

Example Problems

Answer the following questions by examining Figure 15-17 and Figure 15-18.

1. Which letter represents the inferior vena cava?

    answer: a

2. Which letter represents the iliac vein?

    answer: b

3. Which letter represents the femoral vein?

    answer: c

4. Which letter represents the popliteal vein?

    answer: e

5. Which letter represents the anterior tibial vein?

    answer: f

6. Which letter represents the posterior tibial vein?

    answer: g

7. Which letter represents the great saphenous vein?

    answer: d

Work Problems

1. The inferior and superior vena cava transports blood to the _____.

2. The right carotid artery branches off the _____.

3. The right subclavian artery branches off the brachiocephalic artery but the left subclavian artery branches off the _____.

4. Arteries transport blood _____ the heart while veins transport blood _____ the heart.

5. Blood in the popliteal vein will enter the _____ vein next.

6. The radial vein and the ulnar vein join in the _____ region of the arm to form the brachial vein.

7. The anterior tibial vein and the posterior tibial vein will join in the _____ region of the leg to form the popliteal vein.

8. The great saphenous vein joins the _____ vein, which then forms the iliac vein.

9. Which vein is the most lateral; the cephalic vein or the basilic vein?

10. The inferior vena cava must pass through the _____ (a major breathing muscle) before arriving at the right atrium.

Worked Solutions

1. **right atrium**

2. **brachiocephalic artery**

3. **aortic arch**

4. **away from; to**

5. **femoral**

6. **antecubital**

7. **popliteal**

8. **femoral**

9. **cephalic**

10. **diaphragm muscle**

More Blood Vessel Information

Here is some additional information about arteries and veins.

❑ Large blood vessels are made of three layers:

Tunica externa is the outermost layer. It is analogous to the epicardium of the heart.

Tunica interna is the innermost layer. It is analogous to the endocardium of the heart.

Tunica media is the middle layer. It is made of smooth muscle cells. It is analogous to the myocardium of the heart.

❑ Arteries have a thicker tunica media than veins.

❑ Capillaries are vessels that make the transition from arteries to veins.

❑ Capillaries are made of one layer of cells. This thin layer allows for the exchange of oxygen and carbon dioxide.

❑ Veins having a large diameter have one-way valves to assist in getting blood back to the heart.

❑ Oxygenated blood is a bright red color because oxygen is binding to the iron portion of hemoglobin.

❑ Deoxygenated blood is a dark red color because carbon dioxide is binding to the amino acid portion of the hemoglobin molecule.

❑ When you look at the veins in your arm, the blood appears blue. You do not have blue blood. The blood you're looking at is a real dark red color. It only appears blue due to the light waves passing through the layers of skin.

Chapter Problems and Solutions

Problems

1.   When the heart contracts and forces blood to all parts of the body, it is called systole. Which chambers are involved in systole or systolic contraction?

2.   The most inferior portion of the heart is called the _____.

3.   The part of the ECG that represents atrial depolarization is the _____ wave.

4.   The mitral valve is also called the _____ valve.

5.   When the blood arrives at the lungs, it picks up oxygen. What is the name for the thin, one cell layer thick vessels that allow for gas exchange?

6.   Blood in the veins is blue in color and blood in arteries is red in color (true or false).

7.   Blood from the cephalic vein will enter into the basilic vein by passing through which vessel?

8.   The coronary vessels emerge from the base of the _____ to go out to various heart regions.

9.   Many times, if a patient has a heart murmur, it is due to the failure of which valves to close properly?

10.   Relaxation and repolarization of the atria can be identified by which wave of the ECG?

11.   Blood in the iliac veins will enter into the _____.

12.   What two veins will form the superior vena cava?

13.   Blood in the cubital vein will enter the _____.

14.   Blood in the femoral artery will flow into the _____ artery.

15.   Blood in the axillary artery will flow into the _____ artery.

Answers and Solutions

1.   **ventricles**

2.   **apex**

3.   **p**

4.   **bicuspid**

5.   **capillaries**

6.   **false.** All blood is red. Deoxygenated blood is a dark red and only appears blue.

7. **median cubital vein**

8. **ascending aorta**

9. **One of the atrioventricular valves or semilunar valves fails to close properly.**

10. **QRS wave**

11. **inferior vena cava (IVC)**

12. **right and left brachiocephalic veins**

13. **basilic vein**

14. **popliteal**

15. **brachial**

Supplemental Chapter Problems

Problems

1. When we listen to the heart with the use of a stethoscope, we typically hear two sounds. These sounds have been called the lub-dub sounds. The lub (first sound) is the closing of the bicuspid and tricuspid valves. Which valves close thus creating the dub (second sound)?

2. The endocardium of the heart is analogous to the _____ of blood vessels.

3. Blood vessels can dilate and constrict according to environmental conditions because the _____ layer consists of involuntary cells (smooth muscle).

4. Muscle cells of the heart can be identified by the presence of intercalated discs. These cells are located in the _____ layer of the heart.

5. Oxygenated blood can be found on which side of the heart (right or left)?

6. Blood from the right side of the heart is on its way to the _____.

7. Diastole is the term for heart relaxation. What is the term for heart contraction?

8. The two heart sounds created by the heart is due to the closing of _____.

9. If the Purkinje fibers fail to activate, the _____ (which chambers) might not contract.

10. Typically when a person takes a pulse by placing their fingers on the anterior antebrachium, they are feeling the pulsating blood passing through which blood vessel?

11. Blood in the thoracic aorta will flow into the _____ aorta.

12. Blood in the left subclavian artery will flow into the _____ artery.

13. Blood in the right subclavian artery came from the _____ artery.

14. Blood in the basilic vein will enter into the _____.

15. Blood in the radial vein will join with the _____ vein to form the brachial vein.

Answers

1. The closing of the pulmonary and aortic semilunar valves.

2. tunica interna

3. tunica media

4. myocardium

5. left

6. lungs

7. systole

8. valves; atrioventricular valves and semilunar valves

9. ventricles

10. radial artery

11. abdominal

12. left axillary

13. brachiocephalic

14. axillary vein

15. ulnar

Chapter 16
The Lymphatic System

The lymphatic system consists of vessels just like the circulatory system does. These vessels transport interstitial fluid to the veins of the circulatory system. The lymphatic system is also involved in the body's defense mechanism. The lymphatic system consists of a fluid called **lymph,** which is very similar to the plasma of blood.

Lymph and Lymph Vessels

Lymph vessels consist of valves very similar to the valves found in larger veins. These valves are designed to ensure the flow of lymph from the extremities to the larger veins in the upper torso. Lymph from the legs and left side of the body enters into the **thoracic duct,** which then drains into the left subclavian vein. Lymph from the right arm and right side of the head and upper, right torso drains into the **right lymphatic duct,** which then drains into the right subclavian vein.

Lymph Organs

The following bullet list shows you the organs and/or tissues of the lymphatic system and outlines the functions of those organs or tissues:

❑ **Lymph nodes:** Lymph fluid flows through the lymph nodes to be filtered. The lymphocytes in the lymph nodes remove potentially harmful bacteria.

There are numerous lymph nodes scattered throughout the body. Many of the nodes are named according to their location in the body. For example, there are abdominal lymph nodes, thoracic lymph nodes, axillary lymph nodes, and cervical lymph nodes. The lymph nodes basically act as filtering systems for the body. Interstitial fluid enters the lymphatic vessels from various parts of the body and will eventually pass through the lymph nodes. Lymphocytes in the lymph nodes will destroy pathogens such as bacteria. The lymph exiting the nodes is "cleaner" than when it entered. If something should cause blockage in the lymph vessels, tremendous swelling could occur. A classical example is elephantiasis, a swelling of the legs giving the appearance of elephant legs.

❑ **Thymus:** Located posterior to the sternum. Some lymphocytes (one of the five leukocytes) will mature in the thymus gland upon exposure to thymosin. They will then enter into the blood circulation and are called T cells.

The action of the T cells is discussed in the following section.

❑ **Spleen:** Located on the lateral edge of the fundus of the stomach in the left hypochondriac region. The spleen consists of numerous macrophages that will destroy pathogens as they circulate in the blood that is entering the spleen.

The spleen has numerous functions. In addition to destroying pathogens in the blood, the spleen also disposes of dead, worn out, decomposed erythrocytes. There is a lot of blood arriving at the spleen and, therefore, if the spleen ruptured, massive bleeding would occur.

❑ **Tonsils:** There are three sets of tonsils. They are located in the pharynx region. One is located in the nasopharynx region called the **pharyngeal tonsil** (adenoid). There are two located on either side of the uvula (in the oropharynx region) called the **palatine tonsils.** The third set of tonsils is embedded in the posterior edge of the tongue in the oropharynx region and is called the **lingual tonsil.** If the pharyngeal tonsil should swell due to an infection, it would make breathing difficult. If the palatine tonsils should swell, it would make breathing and swallowing difficult. The lymphocytes in the tonsils assist in protecting the entrances to the respiratory and digestive systems.

❑ **Appendix:** The appendix is connected to the first part of the large intestine. It is thought that the appendix is an area for the growth of nonpathogenic bacteria. It appears that the nonpathogenic bacteria keep the pathogenic in check. The appendix also houses a large number of lymphocytes.

The appendix is an extension of the first part of the large intestine called the **cecum.** Its location allows it to provide protection for the large intestine.

❑ **Lymphocytes:** Lymphocytes are produced in the bone and some will mature in the thymus gland. There are several types of lymphocytes. The T cells and B cells are discussed in detail in the following section. The NK cells (natural killer) produce toxic material that directly kills bacteria.

Example Problems

1. Most of the lymphatic fluid will return to venous circulation by entering into which lymphatic duct before entering the left subclavian vein?

    answer: thoracic duct

2. The lymphatic vessels resemble the (arteries or veins) due to the presence of valves inside them.

    answer: veins

3. The thoracic duct drains lymph into the venous circulation from which parts of the body?

    answer: left side of the head, left arm, left thoracic region, and the entire lower body

4. The thymus gland produces _____, which will mature some lymphocytes.

    answer: thymosin

5. Lymphocytes are produced in the bone marrow. However, when some lymphocytes pass through the thymus gland, they are "converted" to what kind of cells?

    answer: T cells

The Lymphatic System and Defense

In order to provide protection, the lymphatic system relies on a special group of white blood cells called lymphocytes. Lymphocytes respond to bacteria, viruses, and toxins in the form of protein molecules produced by some pathogens. Lymphocytes are formed in the bone marrow.

The lymphocytes that travel to the thymus gland and are exposed to thymosin will become T cells and will have a specific function. The lymphocytes that are not exposed to thymosin will become B cells (T for thymus and B for bone). The B cells produce antibodies, and the T cells assist the B cells.

Another group of lymphocytes is known as NK cells. NK cells (natural killer) detect foreign cells and will secrete proteins that will destroy the foreign cell's membrane. Here is a list of the activity of the T and B cells in an effort to defend the body. The process from Step 1 to Step 6 typically takes about 10 to 14 days to accomplish.

1. Bacteria, for example, enter the human body.

2. Macrophages attack the bacteria and ultimately remove the **antigens** from the bacteria and place those antigens on the surface of themselves.

3. The exposure of the antigens on the surface of the macrophages allows T cells to pick up the antigens and transport the antigens to the B cells.

4. Once the B cells receive the antigens, they will analyze the antigens and use the antigen's molecular structure to manufacture specific **antibodies.**

5. The antibodies then bind to the antigens that are still on the bacteria that have not been originally phagocytized.

6. The binding of the antibodies to the antigens (thus creating an antigen/antibody complex) will attract numerous white blood cells. This attraction will bring a lot of phagocytic cells to the infectious site. Eventually, the white blood cells win the battle, and the bacteria die.

While the B cells are in the process of manufacturing antibodies, they also produce memory B cells. Consider the following activity of the memory B cells. The memory B cells are activated when the body is exposed a second time to the same organism (same antigen) as before. This is known as the **second exposure.**

1. Upon exposure a second time (by the same organism and therefore the same antigen) the memory B cells "recognize" that organism with that antigen.

2. The memory B cells begin manufacturing and releasing antibodies that will bind to that antigen. The memory B cells do not need to have the antigen "delivered" to them by the T cells because they "remember" how to make that specific antibody.

3. The antibodies will bind to the antigens.

4. This antigen/antibody complex will attract a host of white blood cells. These white blood cells will destroy the bacteria with that specific antigen/antibody complex.

The process of events in the second exposure takes less time, such as just a few minutes or hours, as compared to the first exposure. Because of this short duration of time, the person with the bacterial invasion doesn't even know it. The body is said to be immune to that particular bacterial organism. Basically, the body is immune to any organism that has that particular antigen.

Example Problems

1. One function of the T cells is to transport _____ to the B cells.

    answer: antigens

2. Antibodies will kill bacteria or other pathogens (true or false).

    answer: false. Antibodies will only bind to the antigens, which will attract other white blood cells. It is the presence of the other white blood cells that kill the pathogen. Don't confuse antibodies with antibiotics.

3. What type of lymphocyte produces antibodies? _____

    answer: B cells

4. Which type of lymphocyte is involved in secondary exposure? _____

    answer: memory B cells

5. Leukocytes are typically attracted to pathogens due to the development of the _____ complex.

    answer: antigen/antibody

Passive and Active Immunity

Active immunity is acquired by being exposed to a pathogenic organism. The lymphocytes respond as described in the preceding section. Think in terms that a person has to actively become exposed to a pathogen to develop this type of immunity.

Another type of active immunity is in the form of a **vaccination.** The patient is given the antigen and the patient's body develops antibodies just as described in the preceding section. The antigen does not cause a disease. The organism causes the disease. Therefore, sometimes just the antigen is given to the patient and other times the entire organism is given, with its antigen still intact, but the disease causing organism is typically weakened or killed.

Passive immunity can also be in the form of a shot given to the patient but in this case, the patient is actually given the antibodies directly. In this situation, the patient does not have to manufacture the antibodies. In this case, however, the immunity is generally short-lived: The antibodies are disposed of by the body, because the body doesn't make them. Figure 16-1 summarizes the information regarding passive and active immunity.

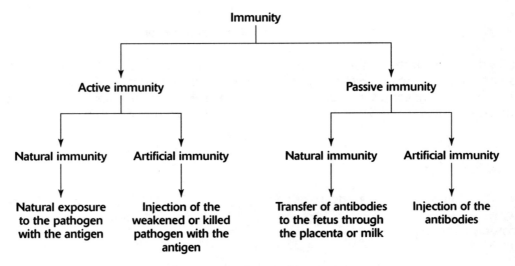

Figure 16-1: Types of immunity.

Example Problems

Identify the type of immunity for the following questions. Use the following terms: active natural, active artificial, passive natural, or passive artificial.

1. When a person is exposed to the chicken pox virus and becomes immune to it, they have developed what type of immunity?

    answer: active natural

2. When a patient has been given a shot that causes their B cells to manufacture antibodies, they have developed what type of immunity?

    answer: active artificial

3. When a patient has been given a shot and the material in the shot binds to the antigens on the pathogens, they have developed what type of immunity?

    answer: passive artificial

4. When a child is born with immunity to a specific disease, they probably developed what type of immunity?

    answer: passive natural

5. A typical flu vaccine is considered to be _____.

    answer: active artificial

Work Problems

1. Lymphocytes are produced in the _____ and some of them will mature in the _____ to become T cells.

2. What is the name of the hormone that is involved in maturing lymphocytes into T cells?

3. Where is the spleen located?

4. What type of lymphocyte is involved in producing antibodies?

5. Which type of lymphocyte produces chemicals, which will directly destroy pathogens?

6. Which of the following will set up a scenario for the attraction of other white blood cells to fight pathogens? (antibiotics or antigen/antibody complexes)

7. T cells will present the _____ to the B cells so the B cells can begin manufacturing antibodies.

8. When people say they are having their adenoids removed, they are actually having the _____ tonsils removed.

9. The thoracic duct drains lymph into which blood vessel?

10. Where is the thymus gland located?

Worked Solutions

1.   **bone marrow; thymus gland**

2.   **thymosin**

3.   **It is located on the left lateral edge of the fundus of the stomach.**

4.   **B cells**

5.   **NK cells (natural killer cells)**

6.   **antigen/antibody complexes**

7.   **antigens**

8.   **pharyngeal**

9.   **left subclavian vein**

10. **It is located posterior to the sternum, superior to the heart.**

Chapter Problems and Solutions

Problems

1.   The fluid that is transported in the lymphatic vessels, which is similar to plasma in the blood, is called _____.

2.   The valves in the lymphatic vessels are shaped in such a manner to ensure that the lymph will flow toward which veins of the body?

3.   The _____ duct collects lymph fluid from the regions inferior to the diaphragm muscle.

4.   How are T cells produced by the thymus gland?

5.   The axillary lymph nodes are located in the _____ region of the body.

6.   Lymph fluid flows through the spleen so the spleen cells can destroy pathogens (true or false).

7.   A person is typically immune to chicken pox because upon the first exposure, he or she produced what kind of lymphocytes?

8.   T cells present the antigens of pathogens to the B cells so the B cells can manufacture antibodies. Where do the T cells get those antigens?

9. The appendix has long been thought to be a vestigial organ with no known function. However, it does appear to protect the entrance to the _____ to harmful pathogens.

10. The tonsils provide protection against pathogens entering the _____ and _____ cavities.

Answers and Solutions

1. **lymph**

2. **left and right subclavian veins**

3. **thoracic**

4. **The thymus gland produces thymosin.** Thymosin converts lymphocytes to T cells.

5. **axillary**

6. **false.** As the blood circulates through the spleen, the cells in the spleen will remove and destroy pathogens. Lymph flows into lymph nodes.

7. **memory B cells**

8. **The T cells get the antigens from the macrophages that originally tried to destroy the pathogens.**

9. **large intestine** (specifically the cecum of the large intestine)

10. **nasal cavity and oral cavity**

Supplemental Chapter Problems

Problems

1. Viruses cause the common cold. However, a person can have a cold many times over. Why are people never immune to the common cold?

2. A patient with breast cancer can have the cancerous breast removed. During the removal of the breast, the surgeon will also remove some axillary lymph nodes. Why is this a common practice?

3. If the thymus gland were to become diminished in its function, which type of lymphocyte would be reduced in numbers?

4. The virus responsible for AIDS seems to "pick on" the T cells. Based on this, why can't our body develop antibodies to begin the process of getting rid of AIDS?

5. There are many times when a child is riding a bicycle and has an accident. The child is then tossed over the handle bars of the bike in such a manner that the spleen is ruptured. Why is it so easy to rupture the spleen?

6. In the body, there are several types of classification for fluid. The intercellular fluid is found within the cells. The interstitial fluid is found between the different tissues of the body. Which of those would become lymph when it enters the lymphatic system?

7. In order to develop immunity to a pathogen, one must be exposed, either actively or passively, to the pathogen's _____.

8. Constant exposure to pathogens trying to enter the body via the nasal cavity or the oral cavity can cause the _____ to swell. This is because the lymphatic tissue inflames due to constant bombardment of pathogens.

9. Thymosin has an affect on which of the five leukocytes?

10. The appendix is attached to and is a part of which portion of the large intestine?

Answers

1. There are numerous viruses that cause the common cold. However, each one has a different antigen. Each time a person is exposed to a new antigen, it takes their body time to develop the B cells and the memory B cells.

2. The lymph in the breast region is drained into the axillary lymph nodes and may contain cancerous cells.

3. T cells

4. In order to manufacture antibodies, the B cells have to receive the antigens from the virus. It is the job of the T cells to transport the antigens to the B cells. If the T cells are inactivated, the B cells will not receive the antigens and cannot manufacture antibodies.

5. It is easy to rupture the spleen because the spleen is located so close to the abdominal wall on the left side.

6. The interstitial fluid enters into the lymphatic system and is then called lymph.

7. antigens

8. tonsils

9. lymphocytes. Upon exposure to thymosin, the lymphocytes will become T cells.

10. cecum. The cecum is the first portion of the large intestine.

Chapter 17
The Respiratory System

Respiration is defined as the exchange of gases between cells and the environment. The gases involved in the human body are oxygen (O_2) and carbon dioxide (CO_2). As the cells of body carry out their daily activity to keep us alive, they use oxygen (burn up oxygen) and generate carbon dioxide. In order to obtain oxygen, we must inhale, and in order to get rid of carbon dioxide, we exhale. The circulatory system plays a vital role with the respiratory system. When you inhale, oxygen goes to the lungs and enters into little sacs called **alveoli.** From the alveoli, the oxygen diffuses into the bloodstream. The bloodstream transports oxygen to all tissues of the body. The carbon dioxide that is generated during cellular activity diffuses into the bloodstream. The blood transports the carbon dioxide to the alveolar sacs of the lungs. You then exhale and the carbon dioxide exits the body.

The best way to study the organs of the respiratory system is to take a molecule of air and follow it as it passes through the system on its way to the cells of the body.

The Respiratory Organs

1. Let's begin with a molecule of air entering into the **nasal cavity.**

2. The air moves to the back of the nose to the area called the **nasopharynx.**

3. Air then moves to the area at the back of the mouth called the **oropharynx.**

4. From the oropharynx, the air enters into the throat region called the **laryngopharynx.**

5. Once in the laryngopharynx, the air encounters two tubes. The anterior tube is the trachea and the posterior tube is the esophagus. Air enters into the **trachea.**

6. Once it passes through the trachea, the air branches into two tubes to enter into each lung. Those two tubes are called the **right and left primary bronchi.**

7. Each primary bronchus branches inside the lungs to form **secondary bronchi.**

8. The secondary bronchi will thus branch to form **tertiary bronchi** and so on.

9. Eventually, the air enters into the final tubes of the respiratory tree called the **bronchioles.**

10. At the terminal ends of the bronchioles are little sacs called **alveoli.**

11. Each alveolus is surrounded by capillaries. Oxygen diffuses through the thin layered wall of the alveolar sacs and into the bloodstream.

12. Oxygen binds to the hemoglobin of the red blood cells. The red blood cells transport the oxygen to the cells of the body.

13. At the cells, the hemoglobin of the red blood cells will drop off oxygen and pick up the waste product, carbon dioxide.

14. The carbon dioxide is transported to the alveolar sacs within the lungs.

15. The carbon dioxide travels in reverse order to exit the body.

The following bullet list gives you some additional information about the respiratory system.

❏ **Nasal cavity:** As air enters the nasal cavity, it encounters **nasal conchae.** The air swirls around the nasal conchae. As the air swirls, it becomes warmer.

❏ **Nasopharynx:** In the nasopharynx region there is a set of tonsils called the **pharyngeal tonsils.** These are commonly called the adenoids.

❏ **Oropharynx:** In the oropharynx region there is a set of tonsils called the **palatine tonsils.**

❏ **Laryngopharynx:** In the laryngopharynx region is one tonsil called the **lingual tonsil.** This tonsil is embedded into the posterior portion of the tongue.

❏ **Trachea:** The opening to the trachea is the **glottis.** There is a piece of cartilage that protects the opening of the glottis called the **epiglottis.** The epiglottis is open when air enters and exits the trachea. The epiglottis closes when food enters the mouth. With the epiglottis closed, the food is basically forced down the esophagus. The trachea consists of cartilage rings. The cartilage rings prevent the trachea from collapsing.

❏ **Bronchi:** The primary bronchi enter into each lung. They branch off the trachea at the point called the **carina.** Because the various bronchi tubes are rather large in diameter, they need cartilage rings or plates of cartilage to prevent collapse.

❏ **Bronchioles:** The bronchioles are quite small in diameter and, therefore, will not collapse. Therefore, they do not have any cartilage. The bronchioles terminate with little sacs called **alveoli.**

❏ **Alveoli:** These small sacs are made of one thin layer of squamous cells. This makes it easy for oxygen and carbon dioxide to pass through to enter the capillaries that surround them. There are about 150 million alveoli per lung.

❏ **Right lung:** The right lung consists of 3 lobes. The right primary bronchus enters the lung at a straighter angle than the left primary bronchus.

❏ **Left lung:** The left lung consists of only 2 lobes. The left primary bronchus enters the lung at a sharper angle than the right primary bronchus. The left lung consists of the **cardiac notch** which supports the apex of the heart.

❏ **Diaphragm muscle:** This is the major breathing muscle. It separates the thoracic cavity from the abdominal cavity. The base of the lungs is nearest the diaphragm (inferior portion of the lungs), and the apex of the lungs is the superior portion of the lungs.

Example Problems

1. Oxygen diffuses from the alveolar sacs to the capillaries or from the capillaries to the alveolar sacs.

   answer: Oxygen diffuses from the alveolar sacs to the capillaries. Carbon dioxide diffuses from the capillaries to the alveolar sacs.

2. The alveolar sacs are located at the ends of which respiratory tubes?

   answer: bronchioles

3. Which respiratory tubes do not have any cartilage supporting them?

   answer: bronchioles

4. What is the name of the structure that prevents food from entering the trachea during swallowing?

    answer: epiglottis

5. The posterior portion of the nasal cavity, which consists of the pharyngeal tonsils, is called the

    answer: nasopharynx

The Process of Inhaling and Exhaling

We are already familiar with the concept that water moves from an area of high concentration to an area of low concentration. Air is very similar in that it moves from an area of high pressure to an area of low pressure. Therefore, in order to get air to enter the lungs, the air pressure in the thoracic cavity has to become less than that of the air pressure outside the thoracic cavity. One way to change the air pressure in the thoracic cavity is to change the size of the thoracic cavity. If a person increased the size of their thoracic cavity, thoracic pressure would go down. If they decreased the size of their thoracic cavity, thoracic pressure would go up. The main muscle that is involved in changing the size of the thoracic cavity is the diaphragm muscle that separates the thoracic cavity from the abdominal cavity.

In order to inhale, the diaphragm muscle goes down. This **increases** the size of the thoracic cavity. This in turn **decreases** the thoracic pressure compared to the outside environment. Air goes into the lungs. In order to exhale, the diaphragm muscle goes up. This **decreases** the size of the thoracic cavity, which in turn **increases** the thoracic pressure compared to the outside environment. Air exits the body.

The rib muscles are also involved in inhalation and exhalation. During the inhalation process, the ribs move upward. This also helps to decrease the internal pressure. During the exhalation process, the ribs move downward. As they move downward they are making the thoracic cavity smaller. This will increase the thoracic pressure and ultimately force air out of the body.

Passive inhalation and exhalation requires a difference of only 2 mm Hg. If atmospheric pressure is 760 mm Hg, the internal thoracic pressure has to drop to 758 mm Hg in order to inhale. In order to exhale, the internal thoracic pressure must increase to 762 mm Hg.

Example Problems

1. What is the name of the major breathing muscle that separates the thoracic cavity from the abdominal cavity?

    answer: diaphragm

2. In order to inhale, the diaphragm muscle moves in which direction?

    answer: down

3. In order to inhale, the size of the thoracic cavity must _____ and the pressure within the thoracic cavity must _____.

    answer: increase; decrease

4. When the diaphragm muscle moves up, it will _____ the size of the thoracic cavity thereby _____ the pressure within the thoracic cavity and air will exit the body.

   answer: decrease; increase

5. What would the internal thoracic pressure have to be in order for inhalation to take place passively if the outside air pressure was 500 mm Hg?

   answer: 498 mm Hg (just 2 mm difference for passive inhalation)

Work Problems

1. The palatine tonsils are located in the _____ region.

2. The glottis is the opening to the _____.

3. The trachea is located (directional term) _____ to the esophagus.

4. What kind of cells make up the epiglottis?

5. Identify the respiratory tubes that enter into each lung.

6. What kind of cells make up the alveoli?

7. The sites where gas exchange occurs within the lungs are the _____.

8. The air pressure within our thoracic cavity will _____ when the thoracic cavity enlarges.

9. The air pressure within our thoracic cavity will _____ when the diaphragm muscle moves downward.

10. When the diaphragm muscle moves downward, it is (contracting or relaxing).

Worked Solutions

1. **oropharynx**

2. **trachea**

3. **anterior**

4. **cartilage**

5. **primary bronchi**

6. **squamous**

7. **alveoli**

8. **decrease**

9. **decrease**

10. **contracting**

The Chloride Shift

During metabolism, the cells of the body produce a significant amount of carbon dioxide (CO_2). Carbon dioxide leaves the cells and enters into the bloodstream. As it enters the bloodstream, it enters the individual erythrocytes. The erythrocytes will transport the carbon dioxide to the alveolar sacs of the lungs to eventually be exhaled. But during the transportation of the carbon dioxide to the lungs, there are a lot of things happening to the carbon dioxide while it is in the erythrocyte. Table 17-1 discusses what is happening to the carbon dioxide in the blood.

The red blood cell consists of water and the plasma around the cells also consists of water. As carbon dioxide is produced, it binds with water. Carbon dioxide plus water will yield **carbonic acid.** Carbonic acid ionizes to form **hydrogen ions** and **bicarbonate ions.** The hydrogen ions have an acidic characteristic. Therefore, the production of hydrogen ions creates a decrease in the blood pH.

Table 17-1 The Fate of Carbon Dioxide in the Bloodstream

Carbon dioxide enters the erythrocyte and will undergo three major activities:		
Carbon dioxide binds to hemoglobin.	Carbon dioxide mixes with water in the cell.	Carbon dioxide mixes with water in the plasma.
23% of the carbon dioxide will bind to hemoglobin.	70% of the carbon dioxide will bind to water in the erythrocyte.	7% of the carbon dioxide will stay in the plasma and bind to water.
This carbon dioxide will ultimately be exhaled.	This combination will form carbonic acid (H_2CO_3).	This combination will form carbonic acid (H_2CO_3).
	Carbonic acid will ionize to form hydrogen ions (H^{1+}) and bicarbonate ions (HCO_3^{1-}).	Carbonic acid will ionize to form hydrogen ions (H^{1+}) and bicarbonate ions (HCO_3^{1-}).
	Hydrogen ions exhibit an acidic characteristic.	Most of the hydrogen ions in the plasma will be used by the stomach (to make stomach acid) or be excreted via the urinary system.
	Therefore, the pH of blood will begin to drop below the standard 7.35–7.45.	
	To prevent developing an acid problem, a lot of the hydrogen ions will bind to hemoglobin.	
	Whenever hydrogen ions bind to something, they no longer behave as an acid.	

(continued)

Table 17-1 The Fate of Carbon Dioxide in the Bloodstream *(continued)*		
	Anything that binds to hydrogen ions is a buffer.	
	Buffers help to stabilize the pH.	

Table 17-1 discusses the normal scenario of hydrogen ions binding to a buffer and therefore the pH is stabilized. Having the right amount of buffers will ultimately create the right pH value. If there are too many buffers, the pH would not be correct just as if there were not enough buffers, the pH would not be correct. If the pH is not what it is supposed to be, enzymes will be affected and, therefore, reduces chemical reactions, which are necessary to maintain homeostasis. Table 17-2 discusses what will happen if the amount of buffers is not correct in the erythrocyte.

Table 17-2 Buffer Action					
Normal Amount of Buffers		***Excessive Amount of Buffers***		***Lack of Buffers***	
Pretend you generated 100 H^{1+} and 100 HCO_3^{1-}. Let's have 85 bicarbonate ions exit the cell (15 still inside the cell).		Pretend you generated 100 H^{1+} and 100 HCO_3^{1-}. Let's have 81 bicarbonate ions exit the cell (19 still inside the cell).		Pretend you generated 100 H^{1+} and 100 HCO_3^{1-}. Let's have 90 bicarbonate ions exit the cell (10 still inside the cell).	
Let's have 80 H^{1+} bind to hemoglobin.	15 H^{1+} will bind to the 15 bicarbonate ions.	Let's have 80 H^{1+} bind to hemoglobin.	19 H^{1+} will bind to the 19 bicarbonate ions.	Let's have 80 H^{1+} bind to hemoglobin.	10 H^{1+} will bind to the 10 bicarbonate ions.
This leaves 5 H^{1+} not bound to a buffer.		This leaves 1 H^{1+} not bound to a buffer.		This leaves 10 H^{1+} not bound to a buffer.	
Pretend 5 H^{1+} will produce a blood pH of 7.35–7.45.		Pretend 1 H^{1+} will produce a blood pH of 8.0.		Pretend 10 H^{1+} will produce a blood pH of 7.0.	
This is perfect homeostasis (the right amount of hydrogen ions).		The blood is too alkaline (fewer hydrogen ions).		The blood is too acidic (too many hydrogen ions).	

In the scenario in Table 17-2, 15 hydrogen ions will bind to 15 bicarbonate ions. The other 85 bicarbonate ions will exit the red blood cell. In this scenario, we are going to pretend that it is essential for 85 bicarbonate ions to exit the cell. In Table 17-2, you can see that any alteration of the amount of bicarbonate ions in the blood will make the blood either too alkaline or too acidic.

In order to get the correct amount of bicarbonate ions to exit the cell, **chloride ions** (from the plasma) will enter the cell. In order to get 85 bicarbonate ions to exit the cell, 85 chloride ions must enter. This is known as the **chloride shift.** If there are too many chloride ions entering the cell, too many bicarbonate ions will exit the cell. This will leave too few bicarbonate ions in the red blood cell. A lack of bicarbonate ions will result in too many hydrogen ions left over. The pH will be too acidic. If there aren't enough chloride ions entering the cell, not enough bicarbonate ions will exit the cell. This will result in too many bicarbonate ions inside the cell. Too many bicarbonate ions will result in a lack of hydrogen ions left over. A lack of hydrogen ions will cause the pH to become too basic (alkaline). The following bullet list summarizes the chloride shift concepts. Keep in mind that bicarbonate ions are buffers.

❑ An **increase** in bicarbonate ions results in a decrease in hydrogen ions, which raises the pH.

❑ An increase in buffers results in a decrease in hydrogen ions, which raises the pH.

❑ A **decrease** in bicarbonate ions results in an increase in hydrogen ions, which lowers the pH.

❑ A decrease in buffers results in an increase in hydrogen ions, which lowers the pH.

Example Problems

1. The concept of the chloride shift is a situation where the _____ ions are entering into a cell and the _____ ions are exiting the cell.

    answer: chloride; bicarbonate

2. If the pH is dropping thus causing the blood to become too acidic, it is due to excess _____ ions.

    answer: hydrogen

3. If there are excessive bicarbonate ions in the red blood cell, there will be (excess or not enough) hydrogen ions in the red blood cell.

    answer: not enough (the bicarbonate ions will buffer too many hydrogen ions)

4. If there are excessive buffers in the red blood cell, there will be (excess or not enough) hydrogen ions in the red blood cell.

    answer: not enough (the bicarbonate ions will buffer too many hydrogen ions)

5. If there are too many buffers in the red blood cell, the pH of the red blood cell will (become too acidic or become too alkaline)?

    answer: too alkaline (if there are too many buffers, they will buffer too many hydrogen ions and the pH will go up)

Work Problems

1. If there is a deficiency in chloride ions, that means there will be fewer chloride ions entering the cell. Therefore, there will be fewer bicarbonate ions exiting the cell. Therefore, there will be (excess or not enough) buffers left inside the red blood cell.

2. Excessive carbon dioxide will ultimately produce excessive _____ ions and therefore cause the pH to drop.

3. The palatine tonsils are located in the _____ region.

4. What ion exits the red blood cell during the chloride shift? _____

5. What is the name of the substance (generic name) that helps to stabilize pH?

6. Cellular activity (metabolism) produces which respiratory gas?

7. An increase in buffers will cause the blood pH to (raise or drop) _____.

8. Carbon dioxide plus water will produce _____.

9. An increase in carbon dioxide will ultimately produce an excessive amount of carbonic acid, which will therefore produce an excess amount of _____ ions and _____ ions.

10. In order for enzymes to function properly (for chemical reactions) the _____ of the blood must be within 7.35–7.45.

Worked Solutions

1. **excess**

2. **hydrogen.** Carbon dioxide binds to water to form carbonic acid. Carbonic acid ionizes to form hydrogen ions. Excess hydrogen ions will result in a drop in the pH.

3. **oropharynx**

4. **bicarbonate ion**

5. **buffers**

6. **carbon dioxide**

7. **raise.** An increase in buffers will result in buffering too many hydrogen ions. This in effect reduces the number of hydrogen ions and therefore the pH goes up (becomes more alkaline).

8. **carbonic acid**

9. **hydrogen; bicarbonate**

10. **pH**

The Control Mechanisms of Breathing

The breathing rate can be altered under a variety of conditions. When a person is exercising, his or her cells need an extra supply of oxygen and nutrients. Therefore, the breathing rate will increase to bring in more oxygen and the heart rate will increase to get the oxygen and nutrients to the cells at a faster rate. Also, because the cells are working harder and faster, they are generating carbon dioxide at a faster rate. Therefore, the breathing rate must increase in order to get rid of the carbon dioxide. The body has several mechanisms that are at work to control the variations in the breathing rate. Table 17-3 describes the various mechanisms.

Table 17-3 Breathing Control Mechanisms	
Receptors	**Action**
Stretch receptors	As the lungs stretch with each breath, stretch receptor cells are stimulated. They send signals, via the vagus nerve, to the medulla oblongata. This triggers the body to begin exhalation.
Pressure receptors	Located in the walls of the ascending aorta and in the internal carotid arteries are pressure receptors. When blood pressure rises or falls, these receptors are activated and will send signals to the medulla oblongata via the vagus and glossopharyngeal nerves. The medulla oblongata will then adjust the heart rate and breathing rate to try to stabilize blood pressure. When blood pressure falls, heart rate increases. When blood pressure rises, heart rate decreases.
Chemical receptors	Chemoreceptors are near the pressure receptors. These receptors are sensitive to the levels of carbon dioxide and oxygen in the blood. These receptors respond more to carbon dioxide than they do oxygen. Therefore, when the carbon dioxide levels vary, the chemoreceptors are activated. When the carbon dioxide levels rise, the chemoreceptors are activated and ultimately causes the patient to breathe faster in an effort to get rid of carbon dioxide. When the carbon dioxide levels fall, the chemoreceptors are activated and ultimately causes the patient to breathe slower in an effort to retain carbon dioxide.

Abnormal Breathing

When a patient's breathing rate slows down, the patient is therefore exhaling less. If they are exhaling less, that means they are exhaling less carbon dioxide. Therefore, they are really retaining more carbon dioxide than normal. If they are retaining more carbon dioxide, that means there is more carbon dioxide to bind to water. Whenever carbon dioxide binds to water, it will indeed produce hydrogen ions. Therefore, it would be safe to say that if there is a high concentration of carbon dioxide, there will be a high concentration of hydrogen ions. Any time the concentration of hydrogen ions goes up, the pH of the solution will go down (becomes more acidic).

When a patient's breathing rate speeds up, the patient is therefore exhaling more. If they are exhaling more, that means they are exhaling more carbon dioxide. Therefore, they are really losing more carbon dioxide than normal. If they are losing more carbon dioxide, that means there is less carbon dioxide to bind to water. If there is less carbon dioxide to bind to water, there will be fewer hydrogen ions formed. Therefore, it would be safe to say that if there is a low concentration of carbon dioxide, there will be a low concentration of hydrogen ions. Any time the concentration of hydrogen ions goes down, the pH of the solution will go up (becomes more alkaline). The following bullet list summarizes this information.

❑ A **decrease** in respiration will result in an increase in carbon dioxide, which will increase the hydrogen ion level thereby lowering the pH.

❑ An **increase** in respiration will result in a decrease in carbon dioxide, which will decrease the hydrogen ion level thereby raising the pH.

When people **hyperventilate,** something has caused them to begin breathing rapidly. They are exhaling too much carbon dioxide. This will ultimately cause the blood pH to become too alkaline. To correct this situation, we typically place a paper bag over a person's mouth and nose. As the hyperventilating person exhales rapidly into the paper bag, he or she is exhaling carbon dioxide, and then re-inhaling this carbon dioxide. This will eventually cause the carbon dioxide level in the body to rise. As the carbon dioxide level rises, the hydrogen ion level will also rise and thus bring the pH back down to a normal level. Hyperventilation will then cease.

Example Problems

1. The stretch receptors are located on the _____.

   answer: lungs

2. When the stretch receptors are stimulated, they will send a signal to the medulla oblongata via the _____ nerve.

   answer: vagus

3. We inhale and exhale about 13 to 15 times per minute. If we were to decrease the respiration to about 9 times per minute, the blood pH will (go up or go down).

   answer: go down (becomes more acidic)

4. If our respiratory rate should increase, the blood pH will (go up or go down).

   answer: go up (becomes more alkaline)

5. In order to stop hyperventilation, the pH must (go up or go down) to return everything to normal.

   answer: go down. When a person is hyperventilating, their blood pH is going up. Therefore, in order to return to normal, the pH needs to come back down.

Protecting the Respiratory System

The respiratory system, just as all other systems, has to function properly in order for the body to maintain homeostasis and to survive in general. Therefore, it is imperative to provide as much protection as possible to the various parts of the respiratory system. The following bullet list discusses the various methods of protecting the respiratory system.

❑ **Rib cage and sternum:** These bony structures provide physical protection for the lungs.

❑ **Goblet cells and ciliated cells:** The trachea and other respiratory tubes are lined with **ciliated columnar cells** and **goblet cells.** The goblet cells produce mucus. This mucus is sticky. Anytime we inhale a foreign particle, this substance gets stuck in the sticky mucus. This prevents the matter from entering into the lungs.

The cilia on the columnar cells wave back and forth in such a manner to propel the mucus upward toward the back of the throat. Once the mucus and the foreign substance arrive at the back of the throat, the body is triggered to cough. We then expel the foreign matter thus protecting the lungs.

❑ **Lung membranes:** Each lung has one membrane surrounding it, which doubles back to form a second layer thereby giving the impression that there are two membranes around each lung.

The layer that is adjacent to the lung tissue is called the **visceral pleura.**

The layer that is nearest the body cavity is called the **parietal pleura.**

❑ **Pleural fluid:** Because the membrane that surrounds the lung doubles back to form the parietal pleura, it actually creates a space between the visceral portion and the parietal portion. There is fluid within this space called **pleural fluid.** This fluid helps to reduce friction as the lungs move with inhalation and exhalation.

❑ **Alveolar macrophages:** These macrophages wander around to devour any potential pathogens that might be present in the alveolar area.

Chapter Problems and Solutions

Problems

1. The two respiratory gases are _____ and _____.

2. What is the name of the cells responsible for moving foreign material away from the lungs in order to protect them?

3. What keeps food from entering the trachea when we swallow?

4. When we say that gas exchange occurs in the lungs, it actually occurs in the _____.

5. What cells produce mucus in the lining of the trachea for the purpose of trapping foreign material?

6. In order to increase the size of the thoracic cavity, the ribs must (go up or go down) and the diaphragm muscle must (go up or go down).

7. In order to inhale, the thoracic pressure must be (more or less) than the atmospheric pressure.

8. What percentage of the carbon dioxide we generate is actually exhaled?

9. What percentage of the carbon dioxide we generate has the potential to produce hydrogen ions in the red blood cell?

10. An increase in carbon dioxide production has the potential to cause the pH to (rise or fall).

Answers and Solutions

1.   **oxygen; carbon dioxide**

2.   **ciliated columnar**

3.   **epiglottis**

4.   **alveoli**

5.   **goblet cells**

6.   **The ribs go up and the diaphragm muscle goes down.**

7.   **less**

8.   **23%**

9.   **70%**

10.   **fall.** The pH will become more acidic.

Supplemental Chapter Problems

Problems

1.   The blood vessels going to the alveoli are (pulmonary venules or pulmonary arterioles).

2.   Emphysema is the destruction of the _____, which makes it very difficult for gas exchange to occur.

3.   Arthritis in the elderly could make inhaling difficult if they have arthritis of the ribs because this could reduce their ability to _____ the size of their thoracic cavity.

4.   When the lungs expand during inhalation, they stimulate the _____ receptors, which will send a signal to the medulla oblongata.

5.   If a person holds their breath, their level of carbon dioxide will _____.

6.   If a person holds their breath, the pH of their blood will (go up or go down).

7.   An increase in carbon dioxide in the blood will _____ the concentration of the hydrogen ions and will ultimately cause the blood to become too _____.

8.   When a person hyperventilates, what is happening to their blood pH?

9. The capillaries going away from the alveoli are (oxygenated or deoxygenated).

10. Pleurisy is a condition where the fluid between the _____ membrane and the _____ of the lungs is infected with either bacteria or viruses.

Answers

1. pulmonary arterioles (remember; arterioles [small arteries] carry blood away from the heart and to the lungs [alveoli])

2. alveoli

3. increase

4. stretch

5. increase

6. go down (become more acidic)

7. increase; acidic

8. Their blood pH is going up (becoming too alkaline).

9. oxygenated

10. visceral; parietal

Chapter 18
The Digestive System

The main function of the digestive system is to take rather large molecules and break them down to form smaller molecules. These small molecules can then enter into the cell where cellular metabolism will take place. In order to break the molecules into smaller molecules, numerous enzymes are required. There are two groups of organs involved in the digestive processes. The major organs of digestion include the mouth, the stomach, and the small intestines. The accessory group of organs include the pancreas, the liver, and the gallbladder.

The Mouth

The process of digestion begins in the mouth. Here are the major features of the mouth:

❑ **Teeth:**

   The teeth are designed to tear and grind the food

   The teeth are identified as; incisors, cuspids, bicuspids, and molars.

❑ **Tongue:** The tongue serves several functions:

   It provides the taste buds as discussed in Chapter 12.

   It serves to push the food to the back of the oral cavity to prepare for swallowing.

   It moves in such a manner to allow proper speech.

❑ **Uvula:** The uvula is the structure that "hangs" down at the back of the throat.

❑ **Salivary glands and salivary amylase:** There are three sets of salivary glands. All three produce the enzyme salivary amylase.

❑ **Digestive enzyme:** Salivary amylase is the digestive enzyme of the mouth and it partially digests carbohydrates.

The food we eat consist primarily of carbohydrates, proteins, and fat. These are all rather large molecules and need to be broken down (digested) in order for the cells to utilize.

Teeth

As food enters the oral cavity, the teeth begin to grind and tear the food. The teeth are identified in Figure 18-1. Adults typically have 28 to 32 teeth. Each jaw consists of 4 incisors, 2 cuspids, 4 bicuspids (premolars), and 6 molars. The last set of molars is also called the wisdom teeth. Some people do not have their molars and therefore have only 28 teeth.

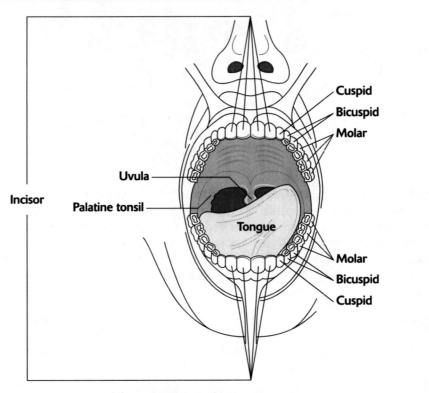

Figure 18-1: Oral structures.

Salivary Glands

There are three sets of salivary glands. The **parotid** glands are located near the masseter muscle.

The **sublingual** glands are located just under the tongue. The **submandibular** glands are located deep in the mandible. All three glands produce salivary amylase. Salivary amylase will begin the digestion of carbohydrates.

The Tongue

The partially digested food is now dissolved and stimulates the various gustatory cells in the taste buds (discussed in Chapter 12). The tongue will push the bolus of food to the back of the throat to prepare for swallowing.

The Uvula

The food will brush against the uvula. A signal is sent to the brain indicating that swallowing is about to take place. The swallowing reflex includes the closing of the epiglottis over the glottis of the trachea to prevent food from going down the trachea. Figure 18-2 shows the action of the epiglottis closing over the trachea.

Figure 18-2a shows the bolus of food being pushed to the back of the mouth in preparation of swallowing. Figure 18-2b shows the bolus of food brushing past the uvula, which then triggers the epiglottis to close over the trachea thereby preventing food from going down the trachea.

The food is eventually swallowed and will enter into the esophagus. The esophagus is made of smooth muscle that contracts in such a manner to create **peristaltic** waves. Due to the action of peristalsis, the partially digested food will move toward the stomach.

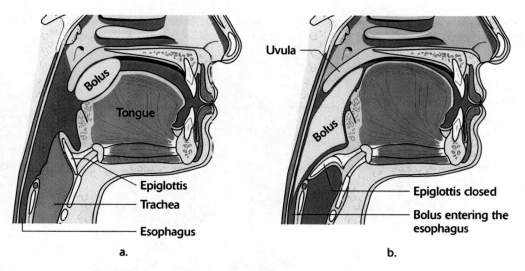

a.   b.

Figure 18-2: The Action of the epiglottis during swallowing.

Example Problems

Use Figure 18-3 and the information discussed regarding the salivary glands to answer
Questions 1 through 3.

1. What is the name for the salivary gland labeled a? _____

   answer: sublingual salivary gland

2. What is the name for the salivary gland labeled b? _____

   answer: submandibular salivary gland

3. What is the name for the salivary gland labeled c? _____

   answer: parotid salivary gland

4. The mouth is responsible for partially digesting what type of food product?

   answer: carbohydrates

5. The epiglottis closes over the _____ to prevent food from going
   down the "wrong tube."

   answer: glottis. The opening to the trachea.

6. The uvula is located (use a directional term) _____to the palatine
   tonsils.

   answer: medial

7. If a person has all of their teeth, they would have _____ incisors, _____
   cuspids, _____ bicuspids, and _____ molars.

   answer: 8 incisors, 4 cuspids, 8 bicuspids, and 12 molars

8. What is the name of the structure in the mouth that is ultimately responsible for getting the epiglottis to close over the glottis of the trachea?

    answer: the uvula

9. Salivary amylase will digest _____ .

    answer: carbohydrates

10. The trachea is located (use a directional term) _____ to the esophagus.

    answer: anterior

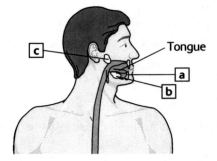

Figure 18-3: The salivary glands.

The Stomach

The bolus of food travels down the esophagus and passes through an opening in the diaphragm muscle to enter into the stomach. This opening is called the **esophageal hiatus.** Figure 18-4 shows this process. The following bullets list select features of the stomach.

❑ **Esophageal sphincter:** The esophagus passes through an opening in the diaphragm called the esophageal hiatus to attach to the stomach. The opening to the stomach is controlled by a smooth muscle called the esophageal sphincter.

❑ **Gastric rugae:** The gastric rugae are the muscular ridges inside the stomach. The rugae will flatten when food enters the stomach so the stomach can expand.

❑ **Hydrochloric acid and pepsinogen:** HCl and pepsinogen are produced by the stomach cells. The combination of HCl and pepsinogen will produce the digestive enzyme, **pepsin.**

❑ **Pyloric sphincter:** The pyloric sphincter is a smooth muscle at the distal end of the stomach. Once the partially digested food passes through this sphincter it is on its way to the small intestine.

❑ **Digestive enzyme:** Pepsin is the digestive enzyme of the stomach and it partially digests protein.

The bolus of food entering the stomach still consists of protein and fat and partially digested carbohydrates.

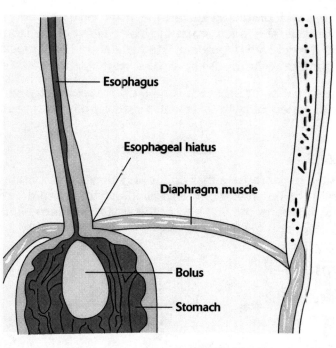

Figure 18-4: The esophageal hiatus.

The Esophageal Sphincter

The esophageal sphincter is a circular muscle that opens and closes involuntarily at the entrance to the stomach. When food is in the stomach, the esophageal sphincter closes to contain the food and the digestive juices. There are other stomach muscles that will cause the stomach to contract and relax in such a manner to get the food thoroughly mixed with gastric juices.

The Gastric Rugae

As food enters the stomach, the stomach will stretch. The rugae allow the stomach to have the ability to stretch. As the stomach stretches, a hormone is released from stomach cells, which will cause other stomach cells to begin producing hydrochloric acid and pepsinogen.

Hydrochloric Acid, Pepsinogen, Pepsin

As the stomach stretches, it produces **hydrochloric acid** and **pepsinogen.** The combination of hydrochloric acid and pepsinogen yields the digestive enzyme, pepsin. Pepsin digests only protein. Hydrochloric acid not only activates pepsinogen to form pepsin, but it also serves to destroy pathogens. As we eat, pathogens can easily enter into the digestive system. Hydrochloric acid creates an environment in the stomach with a pH of 1 to 2. This extreme acid level will kill most pathogens.

If the esophageal sphincter should open while food is in the stomach, the movement of the stomach could cause some of the acid to slosh up into the esophagus. The esophagus does not have any protection against acid. The acid then causes a burning sensation in the esophagus. This burning sensation is described as heartburn. Because the heart is not involved at all, a more appropriate term is **esophageal reflux** or **acid reflux.**

The lining of the stomach is well protected against the acid it produces. The stomach has cells that produce a thick layer of mucus. The mucus coats the inside lining of the stomach and, therefore, buffers the acid. If any of the cells fail to produce adequate amounts of mucus, the acid can begin to destroy the lining of the stomach. This is known as a **stomach ulcer.**

The partially digested material in the stomach is now called **acidic chyme.** The acidic chyme consists of fat, partially digested carbohydrates, and partially digested protein.

Pyloric Sphincter

The pyloric sphincter is a circular muscle that opens and closes involuntarily at the distal end of the stomach leading to the beginning of the small intestine. When the pyloric sphincter opens, the acidic chyme will enter the first part of the small intestine called the **duodenum.**

Example Problems

1. Hydrochloric acid converts _____ to the active form called pepsin.

   answer: pepsinogen

2. The stomach mainly digests _____.

   answer: protein

3. The muscular ridges on the inside of the stomach that allow the stomach to stretch when food is present are called _____.

   answer: rugae or gastric rugae

4. When food is in the mouth it is called a bolus. But, when it enters the stomach and mixes with acid it is called acidic _____.

   answer: chyme

5. The esophageal hiatus is an opening in the _____, which allows the esophagus to pass to enter into the stomach.

   answer: diaphragm

The Small Intestine

The following bullet list gives you select features of the small intestine. Due to the numerous enzymes, the small intestine does more digestion than the stomach. The small intestine is considerably longer than the stomach and therefore food spends more time in the small intestine than it does in the stomach. Spending more time in the small intestine allows for more digestion. So, the combination of more enzymes and longer time makes for more digestion in the small intestine than in the stomach.

- ❑ **Duodenum:** This is the first part of the small intestine. It is about 1 foot in length.

- ❑ **Hepatopancreatic sphincter:** The pancreas produces numerous enzymes. These enzymes will leave the pancreas and enter into the duodenum at the site of the hepatopancreatic sphincter. Bile from the gallbladder will also enter the duodenum at the same site.

❑ **Jejunum:** This is the middle portion of the small intestine. It is about 8 feet in length and consists of numerous structures called **villi.** Most of the digestion and absorption occurs in this region.

❑ **Villi:** These are structures lining the inside of the small intestine. Digested material is absorbed through the villi and enters into the bloodstream.

❑ **Ileum:** This is the last segment of the small intestine and is about 10 feet in length. The ileum consists of numerous villi. Any material not absorbed will leave the ileum and enter into the first part of the large intestine called the **cecum.**

❑ **Digestive enzymes:** The small intestine produces numerous enzymes for the purpose of digestion.

> Maltase will digest maltose, a carbohydrate.
>
> Sucrase will digest sucrose, a carbohydrate.
>
> Lactase will digest lactose, a carbohydrate.
>
> Peptidase will digest protein.

The Duodenum

The duodenum is the first part of the small intestine. The acidic chyme from the stomach enters into the duodenum. Therefore, the duodenum is exposed to acid. The duodenum does not have the mucus protection that the stomach has. If the duodenum is not protected, it is quite likely that an ulcer could result. The ulcers in the duodenum are called **duodenal ulcers.** However, the duodenum does have some protection. The pancreas produces buffers. The buffers leave the pancreas and enter the duodenum by passing through the hepatopancreatic sphincter. Figure 18-5 shows how buffers leave the pancreas and enter into the duodenum.

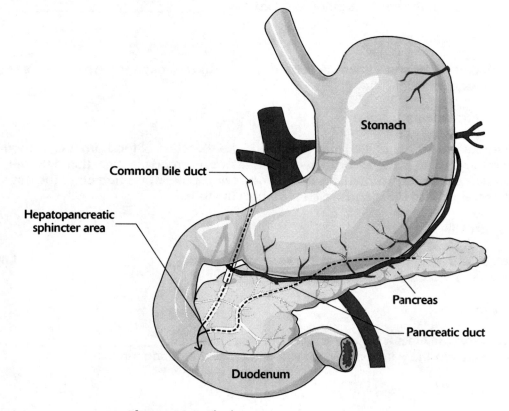

Figure 18-5: The hepatopancreatic sphincter.

The Jejunum

Most of the digestion occurs in the second part of the small intestine. The cells lining the small intestine produce numerous enzymes that will continue the digestion process. The acidic chyme that has entered the small intestine consists of partially digested carbohydrates, partially digest protein, and fat. The enzymes maltase, sucrase, and lactase will continue the digestion of carbohydrates. The peptidase that is also produced by the cells of the small intestine will continue the digestion of protein.

There are numerous enzymes that are produced in the pancreas that will enter into the duodenum and will digest food in the jejunum as well. Figure 18-5 shows how the enzymes that are produced in the pancreas will enter into the duodenum. The pancreas produces numerous enzymes that will help to complete the process of digestion in the small intestine.

Once the food is thoroughly digested it has to be absorbed through the walls of the small intestine to enter into the bloodstream so it can be transported to the cells of the body. The structures that permit maximum absorption are called **villi.**

The Ileum

The ileum is the last segment of the small intestine. It is in this segment that final absorption takes place. Any digested material that is not absorbed will enter into the large intestine and will become waste. At the junction of the first part of the large intestine, called the **cecum,** is an involuntary muscle called the **ileocecal sphincter.** When this sphincter opens, the material in the ileum will enter into the cecum and is now in the large intestine as waste products.

Example Problems

1. The first part of the small intestine is called the _____.

    answer: duodenum

2. Which part of the small intestine is connected to the stomach and which part is connected to the large intestine?

    answer: duodenum; ileum

3. An important aspect of the small intestine is the digestion of food products. Another important aspect is the absorption of the nutrients (formed via digestion) into the bloodstream so the various cells of the body can receive those nutrients. The nutrients are absorbed into the bloodstream via what structures?

    answer: villi

4. The pH of the acid coming from the stomach is about _____ and the pH in the small intestine is about _____.

    answer: 1 to 2; 7 to 8

5. In order to maintain a pH of 7 to 8 in the small intestine, _____ are released from the pancreas, which will enter into the duodenum.

    answer: buffers

Work Problems

1. The small intestine is made of three major regions. Identify those regions in correct sequence regarding the movement of food through the small intestine.

2. What is the name of the muscular action that moves the food through the esophagus, through the stomach, and through the small intestine?

3. The largest salivary gland is the parotid gland and it is located near the _____ muscle.

4. When the _____ cells of the taste buds are stimulated by dissolved food, they will send signals to the brain for the interpretation of flavor.

5. When we begin to swallow, the _____ will close over the _____ to prevent choking.

6. The stomach produces _____ enzyme and the small intestine produces _____ enzymes. (State your answer in numerals.)

7. Food that is not digested or absorbed in the small intestine will be considered waste and will enter the first part of the large intestine called the _____.

8. Many people discuss canine teeth. Which teeth are considered to be (in laymen's terms) the canine teeth?

9. The stomach is made of the cardia region, the fundus region, and the pylorus region. Which of those regions attach to the duodenum of the small intestine?

10. The enzyme in the mouth mainly digests _____.

Worked Solutions

1. **duodenum, jejunum, and ileum**

2. **peristalsis**

3. **masseter**

4. **gustatory cells** (refer to Chapter 12)

5. **epiglottis; trachea (or opening to the trachea called the glottis)**

6. **1 (pepsin); 4 (maltase, sucrase, lactase, peptidase)**

7. **cecum**

8. **cuspids**

9. **pylorus**

10. **carbohydrates**

The Pancreas

There are numerous enzymes produced by the pancreas. Because food does not enter into the pancreas to be digested, it is considered an accessory organ of digestion. Consider the following select features of the pancreas.

❑ **Buffers:** Buffers from the pancreas will travel in the pancreatic duct and through the hepatopancreatic sphincter into the duodenum.

❑ **Digestive enzymes:** The pancreas produces numerous enzymes that will enter into the small intestine to digest food.

Trypsin will digest protein.

Chymotrypsin will digest protein.

Carboxypeptidase will digest protein.

Pancreatic amylase will digest carbohydrates.

Lipase will digest fat.

❑ **Hepatopancreatic sphincter:** This sphincter controls the flow of material into the duodenum from the pancreatic duct of the pancreas and the common bile duct of the liver.

Buffers

When partially digested chyme enters the small intestine, it is rather acidic due to the acid nature of the gastric juices of the stomach. The pH of the acid from the stomach is 1 to 2 and the pH of the small intestine is 7 to 8. In order for the enzymes in the small intestine to function properly, the pH has to be 7 to 8. If acid is entering into the duodenum from the stomach, the small intestine pH could decrease. If the pH decreases too much, the enzymes will denature and therefore not do an adequate job of digesting food. Therefore, to stabilize the pH (keep it at 7 to 8), the pancreas releases buffers.

The acid entering into the small intestine could also cause ulcers in the duodenum called **duodenal ulcers.** The duodenum does not have much protection against acid. The act of stabilizing the pH will protect against duodenal ulcers.

Digestive Enzymes from the Pancreas

When acidic chyme consisting of carbohydrates, protein, and fat enter into the small intestine, the pancreas is stimulated to release a host of digestive enzymes. The small intestine itself produces **carbohydrases** such as maltase, sucrase, and lactase. The pancreas will add its carbohydrase, pancreatic amylase. Carbohydrates are, therefore, thoroughly digested in the small intestine.

The small intestine itself produces a **protease** called peptidase. The pancreas will add its proteases such as; trypsin, chymotrypsin, and carboxypeptidase. Proteins are therefore thoroughly digested in the small intestine.

The small intestine itself does not produce an enzyme to digest fat. When acidic chyme with fat enters the small intestine, the pancreas is stimulated to release **lipases,** which will digest lipids such as fat. Fat is finally digested in the small intestine but with an enzyme from the pancreas.

Figure 18-5 shows the pancreatic enzymes traveling through the pancreatic duct toward the duodenum. The enzymes will pass through an opening called the **hepatopancreatic sphincter.** Buffers will also pass through the hepatopancreatic sphincter.

The dotted line in the pancreatic duct represents the flow of digestive enzymes and buffers, produced by the pancreas, to the duodenum. These products will enter the duodenum by passing through the hepatopancreatic sphincter. Notice, bile from the liver and gallbladder will pass through the common bile duct and will also enter the duodenum at the site of the hepatopancreatic sphincter.

Fat is a very difficult substance to break down. Lipase can digest the fat but it does a more efficient job with the help of bile. **Bile** is produced in the liver and stored in the gallbladder. When fat enters the small intestine, the gallbladder releases its stored bile into the small intestine. Bile will emulsify the fat. Once the fat is emulsified, lipase will begin to work on it. Emulsified fat is easier to digest than fat that has not been emulsified.

The Liver and Gallbladder

The following bullet list highlights the features of the liver and gallbladder. The liver produces bile and the gallbladder stores the bile. When bile is needed, the hepatopancreatic sphincter opens and bile will enter the duodenum of the small intestine. Bile is then used to emulsify fat. Bile does not digest fat. Refer to Figure 18-5 and examine Figure 18-6 to see how bile enters the duodenum.

❑ The liver consists of hepatocytes. Hepatocytes produce bile.

❑ Tubes associated with the liver:

> The liver consists of the left and right hepatic ducts.

> The left and right hepatic ducts join to form the common hepatic duct.

> The common hepatic duct forms the common bile duct.

> The common bile duct leads to the hepatopancreatic sphincter, which opens into the duodenum.

❑ Tubes associated with the gallbladder:

> Bile leaves the gallbladder by traveling through the cystic duct.

> The cystic duct joins the common bile duct.

> The common bile duct leads to the hepatopancreatic sphincter, which opens into the duodenum.

❑ Emulsification is the process of taking large "globs" of fat and breaking them down to form small "globs" of fat. It is easier for lipase to digest small globs of fat than large globs.

The liver produces bile and bile is stored in the gallbladder. As bile leaves the liver, it can enter the cystic duct to go to the gallbladder for storage.

Example Problems

Use Figure 18-6 and the information in preceding bullet list to answer the following questions.

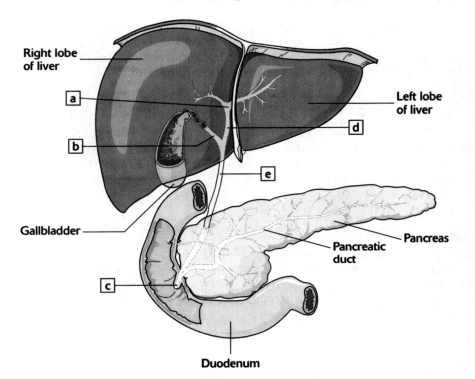

Figure 18-6: The liver and gallbladder.

1. What is the name of structure a in Figure 18-6?

    answer: left and right hepatic ducts

2. What is the name of structure b in Figure 18-6?

    answer: cystic duct

3. What is the name of structure c in Figure 18-6?

    answer: hepatopancreatic sphincter

4. What is the name of structure d in Figure 18-6?

    answer: common hepatic duct

5. What is the name of structure e in Figure 18-6?

    answer: common bile duct

Work Problems

1. The hepatopancreatic sphincter is joined by the tube from the pancreas and the tube from the gallbladder and liver. What are those two tubes called?

2. When the hepatopancreatic sphincter opens, bile and pancreatic enzymes will enter into the _____ of the small intestine.

3. The pancreas produces (how many) _____ enzymes that will enter into the small intestine to digest food.

4. After the digestive enzymes have entered into the small intestine from the pancreas and are added to the digestive enzymes already produced in the small intestine, there will be a total of how many enzymes at work in the small intestine?

5. In order for fat to be digested in the small intestine, the pancreas must release the enzyme _____ into the duodenum.

6. Bile is produced in the _____ and is stored in the _____.

7. _____ will emulsify fat so the enzyme, _____ can do a more efficient job of digesting the fat.

8. What is the name of the tube that drains bile from the gallbladder into the common bile duct?

9. Because the liver, gallbladder, and pancreas are heavily involved in the digestive processes (even though food does not enter those organs); they are called _____ organs of digestion.

10. In order for the enzymes from the pancreas and the enzymes produced in the small intestine to work properly, the pH must be _____.

Worked Solutions

1. **pancreatic duct and common bile duct**

2. **duodenum**

3. **5:** trypsin, chymotrypsin, carboxypeptidase, pancreatic amylase, and lipase

4. **9:** 4 already produced in the small intestine and 5 from the pancreas

5. **lipase**

6. **liver; gallbladder**

7. **Bile; lipase**

8. **cystic duct**

9. **accessory**

10. **7 to 8**

The Large Intestine

Any food material that is not digested or is not absorbed by the villi of the small intestine will pass through the **ileocecal sphincter** and enter into the first part of the large intestine called the **cecum.** The large intestine does not digest food nor does it absorb nutrients. Its main function regarding the digestive system is to absorb water back into the bloodstream to prevent dehydration. Consider the following select features of the large intestine (and see Figure 18-7).

❑ **Cecum:** This is the first part of the large intestine.

❑ **Appendix:** The vermiform appendix is attached to the cecum. It averages 7–9 cm in length.

❑ **Ascending colon:** Due to peristaltic action of the smooth muscles of the large intestine, waste material is moved through the ascending colon. This portion is located on the right side of the body.

❑ **Hepatic flexure:** This is a bend in the large intestine located in right hypochondriac region. It's called the hepatic flexure because it is a bend located near the liver (hepatic).

❑ **Transverse colon:** This portion of the large intestine is located in a transverse position in the abdominal cavity. Due to peristaltic action, waste in this portion moves from the right side to the left side of the body.

❑ **Splenic flexure:** This is the second bend in the large intestine and is located in the left hypochondriac region. It is called the splenic flexure because it is a bend located near the spleen (splenic).

❑ **Descending colon:** Due to peristaltic action, waste material is moved through the descending colon. This portion is located on the left side of the body.

❑ **Sigmoid flexure:** This is the third bend in the large intestine located in the left iliac region. It is called the sigmoid flexure because it leads to the S-shaped sigmoid colon.

❑ **Sigmoid colon:** The sigmoid colon curves from the left side of the body to the midline of the body.

❑ **Rectum:** The sigmoid colon empties waste material into the rectum. From the rectum, the waste passes through the anus.

The waste products that enter into the large intestine consist of approximately 2L of water. It is imperative that some of this water is reabsorbed into the bloodstream. As the waste material is traveling through the large intestine (via peristaltic action of smooth muscles) water is reabsorbed into the nearby blood vessels. If the waste material passes through the large intestine too slow, excess water is removed from the waste and constipation occurs. If the waste material passes through the large intestine too fast, very little water is reabsorbed. Therefore, the waste material is very watery and diarrhea occurs.

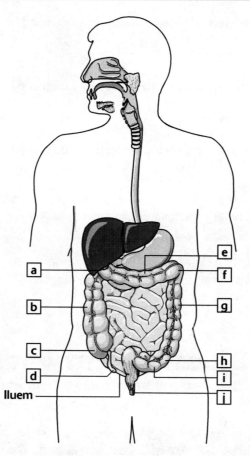

Figure 18-7: The large intestine.

Example Problems

Use Figure 18-7 and the information in the preceding section to answer Questions 1 through 10.

1. What is the name for the large intestine structure labeled a?

   answer: hepatic flexure

2. What is the name for the large intestine structure labeled b?

   answer: ascending colon

3. What is the name for the large intestine structure labeled c?

   answer: cecum

4. What is the name for the large intestine structure labeled d?

   answer: appendix

5. What is the name for the large intestine structure labeled e?

   answer: transverse colon

6. What is the name for the large intestine structure labeled f?

    answer: splenic flexure

7. What is the name for the large intestine structure labeled g?

    answer: descending colon

8. What is the name for the large intestine structure labeled h?

    answer: sigmoid flexure

9. What is the name for the large intestine structure labeled i?

    answer: sigmoid colon

10. What is the name for the large intestine structure labeled j?

    answer: rectum

Summary of Digestive Enzymes

Table 18-1 lists the organs of digestion and summarizes the enzymes that are produced and/or used in that organ.

Table 18-1 Enzymes of the Digestive System		
Organ of Digestion	*Enzyme*	*Function*
Mouth	Salivary amylase	Partially digests carbohydrates
Stomach	Pepsin	Partially digests protein
Small intestine	Maltase	Digests carbohydrates
	Sucrase	Digests carbohydrates
	Lactase	Digests carbohydrates
	Peptidase	Digests protein
	Trypsin (from the pancreas)	Digests protein
	Chymotrypsin (from the pancreas)	Digests protein
	Carboxypeptidase (from the pancreas)	Digests protein
	Pancreatic amylase (from the pancreas)	Digests carbohydrates
	Lipase (from the pancreas)	Digests fat

Hormones of the Digestive System

In order to accomplish the task of digestion, there has to be numerous digestive enzymes. Those enzymes will only work at optimum pH. In this case the pH is 7 to 8. The pH is stabilized by the action of the buffers produced by the pancreas, which enter the duodenum. In addition, in order to get the enzymes released from the pancreas and bile to be released from the gallbladder; hormones are involved. Consider the following list of select hormones and their function.

❑ **Gastrin:**

When the stomach stretches due to the presence of food, stomach cells release gastrin.

Gastrin targets other cells of the stomach.

Gastrin causes those cells to produce HCl and pepsinogen.

The combination of HCl and pepsinogen will yield pepsin.

❑ **Cholecystokinin:**

Due to the presence of acidic chyme consisting of protein and fat, the cells of the small-intestine will release cholecystokinin (CCK).

CCK targets the gallbladder and the pancreas.

CCK causes the gallbladder to release bile.

CCK causes the pancreas to release its digestive enzymes (collectively called **pancreatin**).

CCK also targets the hepatopancreatic sphincter to get it to open to allow pancreatic enzymes and buffers and bile to enter the duodenum.

❑ **Secretin:**

Due to the presence of acidic chyme, the cells of the small intestine will release secretin.

Secretin targets the liver and the pancreas.

Secretin causes the liver to begin making bile.

Secretin causes the pancreas to release buffers.

Work Problems

1. Name the hormone that causes the gallbladder to release bile.

2. Name the hormone that causes the release of buffers from the pancreas.

3. What causes the small intestine to release cholecystokinin?

4. True or false. The large intestine will complete the digestion of fat.

5. Secretin does not stabilize the pH in the small intestine but it is necessary in order for the pH to become stabilized. How is secretin necessary for this process to work?

6. The activation of (or conversion of) _____ produces the digestive enzyme pepsin.

7. A patient with pancreatic cancer will likely have difficulty in producing various hormones but will also have a difficult time in producing _____

8. Based on the information in Table 18-1, which organ does the most digestion?

9. Gastrin is released by stomach cells and will target _____.

10. What hormone causes the hepatopancreatic sphincter to open?

Worked Solutions

1. **cholecystokinin**

2. **secretin**

3. **The presence of fat and protein will cause the release of CCK.**

4. **false.** The large intestine does not digest food.

5. **Secretin causes the pancreas to release buffers.** Buffers will stabilize the pH of the small intestine.

6. **pepsinogen**

7. **digestive enzymes**

8. **small intestine**

9. **stomach cells**

10. **cholecystokinin (CCK)**

Chapter Problems and Solutions

Problems

1. True or false: The stomach does the most digestion.

2. What hormone is required to get the pancreas to release its digestive enzymes into the duodenum?

3. What hormone is required to get the pancreas to release its buffers into the duodenum?

4. The stomach has a lot of mucus, which will provide protection for the stomach lining against the acid that is produced by the stomach. The small intestine does not have this type of protection. What provides protection against acid in the small intestine?

5. When the small intestine has a lot of _____ in the chyme, there will be excess bile released into the duodenum.

6. The hormones cholecystokinin and secretin are produced by the cells in the _____.

7. The main job of the large intestine is to reabsorb _____ back into the bloodstream.

8. Name the enzymes that are involved in digesting carbohydrates associated with the entire digestive system.

9. Name the enzymes that are involved in digesting protein associated with the entire digestive system.

10. Name the enzyme that is involved in digesting fat associated with the entire digestive system.

Answers and Solutions

1. **false.** The small intestine digests more than the stomach.

2. **cholecystokinin (CCK)**

3. **secretin**

4. **buffers (from the pancreas) will enter into the small intestine to protect it against acid.**

5. **fat.** Bile emulsifies fat.

6. **small intestine**

7. **water**

8. **salivary amylase, maltase, sucrase, lactase, pancreatic amylase**

9. **pepsin, peptidase, trypsin, chymotrypsin, carboxypeptidase**

10. **lipase**

Supplemental Chapter Problems

Problems

1. When the _____ sphincter fails to close properly, some stomach contents can enter into the esophagus. This is known as esophageal reflux.

2. The ileum is a part of the _____ and the ilium is a part of the _____.

3. Heartburn is a condition where stomach acid is "sloshing" back into the esophagus. The acid begins to burn the inside lining of the esophagus at about the height of the location of the heart. It does not affect the heart. Therefore, a more correct term to describe this condition is _____.

4. Rapid peristalsis in the large intestine could result in what medical condition?

5. Lactose intolerance is a condition where the patient is not producing enough of what enzyme?

6. A patient with gallbladder problems may have difficulty digesting which type of food substance?

7. A decrease in buffers coming from the pancreas could result in the change of the pH of the small intestine. In this situation, the pH of the small intestine will (rise or fall).

8. A lack of secretin could result in a (rise or fall) in the pH of the small intestine.

9. When food enters the stomach, it causes the stomach to stretch, thereby releasing gastrin. Stress also can cause the stomach to release gastrin. Based on this and the information on the hormones in the digestive system, explain why stress may cause ulcers.

10. A hiatal hernia is the inflammation of which structure discussed in this chapter? The name of the condition is a hint.

Answers

1. esophageal

2. small intestine; os coxae (hip structure). Notice the difference in the spelling of ileum and ilium.

3. esophageal reflux or acid reflux

4. diarrhea

5. lactase

6. fat. Gallbladder problems may result in a decrease in bile storage and release. Therefore, fat cannot be emulsified, which makes it very difficult to digest the fat.

7. fall (becomes more acidic)

8. fall. A lack of secretin could result in a lack of buffer release from the pancreas.

9. Regardless of what causes the release of gastrin, gastrin will cause the stomach cells to release HCl. Hydrochloric acid causes the pH of the stomach to reach a value of 1 to 2. Excess acid production, without the presence of food, could create ulcers.

10. The esophageal hiatus in the diaphragm muscle.

Chapter 19
Metabolism

After food is ingested, your digestive system begins to break the food down into its molecular form, which is referred to as **nutrients.** These nutrients are then absorbed from the small intestine into the bloodstream. The nutrients then travel to all cells of the body. The cells then absorb the nutrients. The cells will then metabolize the nutrients. Metabolism is defined as the sum total of all the chemical reactions in the body that are necessary to maintain homeostasis. Metabolism involves catabolic reactions **(catabolism)** and anabolic reactions **(anabolism).**

Catabolism

Catabolism is the breakdown of large molecules into smaller molecules. Catabolism usually occurs in two stages; the first in the cytosol of the cell and the second in the mitochondria of the cell. Consider the following information regarding catabolism in the cytosol:

❑ Large organic molecules (from digestion) enter into the cytosol of the cell.

❑ There are enzymes in the cell that will convert those large organic molecules into smaller molecules.

❑ During the conversion of these molecules, a small amount of ATP (adenosine triphosphate) is released.

❑ Catabolism generally involves the conversion of the following food types:

When fats are digested, they form fatty acids, which can enter into the mitochondria.

When carbohydrates are digested, they form glucose, which will then form pyruvic acid, which can enter into the mitochondria.

When proteins are digested, they form amino acids, which will then form pyruvic acid, which can enter into the mitochondria.

In the mitochondria, catabolism behaves as follows:

❑ Fatty acids and pyruvic acid will enter into the mitochondria.

❑ There are enzymes in the **mitochondria** that will convert the various organic molecules into a large amount of ATP.

Anabolism

The ATP that is generated during catabolism can be used for numerous activities in the body. One such activity is the building of molecules necessary to carry out various functions. This building process is called anabolism. Consider the following select anabolic activities of the cell.

❑ Glucose enters into the liver and the liver cells will add numerous glucose molecules together to produce a larger molecule called glycogen. Glycogen can then be stored for later use.

❑ Amino acids are bonded together to produce various types of protein in the body. One such protein structure would be hemoglobin.

❑ A variety of other molecules are bonded together to form cell membrane molecules.

Example Problems

1. Where in the cell is the most ATP formed?

    answer: In the mitochondria.

2. Metabolism is comprised of two cellular activities. They are _____ and _____.

    answer: anabolism and catabolism

Use one of these two terms to answer the following questions: catabolism or anabolism.

3. Pyruvic acid is formed during the process of _____.

    answer: catabolism

4. The glycogen that is stored in liver cells was formed during the process of _____.

    answer: anabolism

5. ATP is generated during the process of _____.

    answer: catabolism

Metabolism of Carbohydrates

The majority of the food we eat consists of carbohydrates, protein, and fat. Carbohydrates will be digested to form glucose. Glucose is then used in catabolic reactions. Protein will be digested to form amino acids and amino acids will be used in catabolic reactions. Fats will be digested to form fatty acids and glycerol, which are then used in catabolic reactions. The following bullets discuss the metabolism of carbohydrates (Figure 19-1 illustrates the metabolism of carbohydrates).

Catabolism:

❑ Carbohydrates (such as starch found in potatoes and bread) will be digested to form glucose.

❑ Glucose will enter into the cell and is converted to glucose-6-phosphate.

❏ Glucose-6-phosphate will then convert to phosphoglyceraldehyde (PGAL).

❏ Phosphoglyceraldehyde will convert to form pyruvic acid.

❏ The conversion of glucose to ultimately pyruvic acid in the cytosol is known as **glycolysis.**

❏ Pyruvic acid will enter into the mitochondria.

❏ There are numerous reactions occurring in the mitochondria that lead to the formation of ATP.

❏ The chemical reactions in the mitochondria that lead to the formation of ATP are known as **the Kreb's reactions** (named for Johannes Kreb), and the **electron transport system.**

Anabolism:

❏ Any glucose molecules that are not needed immediately for the production of ATP will be bonded together to form glycogen. Glycogen is then stored in liver cells and skeletal muscle cells.

❏ Glycogen can then undergo catabolism to "reform" glucose at a later time.

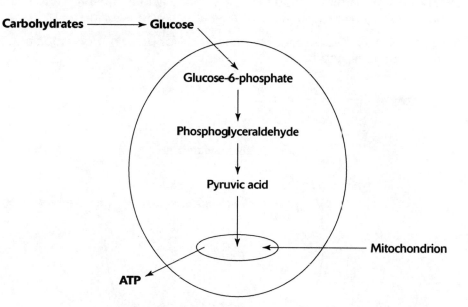

Figure 19-1: Carbohydrate metabolism.

Metabolism of Protein

There are hundreds of thousands of different proteins in the body. Each protein is made of various amino acids. The body uses only 20 different kinds of amino acids even though some proteins can consist of 574 amino acids. The 20 different amino acids are bonded together in a variety of sequences to create the many different protein molecules. The limited number of amino acids the body uses to create thousands of different protein molecules is analogous to the English alphabet. There are only 26 letters in the alphabet but yet there are thousands of different words. The letters are just placed in different sequences to create different words. The sequence of the amino acids is determined by the DNA molecule in the nucleus of the cell. Figure 19-2 illustrates the metabolism of protein, while the following bullet list discusses the metabolism of protein.

Catabolism:

❑ Protein is digested to form amino acids.

❑ Amino acids can enter into the cell and is broken down.

❑ A portion of the amino acid molecule is called the amino group. During catabolism, this amino group is removed thus forming ammonia.

❑ Liver cells will convert the ammonia molecules to a less toxic molecule called **urea.**

❑ Urea is then excreted in the urine.

❑ Also, when amino acids enter into the cell, they can convert (ultimately) to pyruvic acid.

❑ Pyruvic acid will enter into the mitochondria.

❑ There are numerous reactions occurring in the mitochondria that lead to the formation of ATP.

Anabolism:

❑ During anabolism, the amino group from one amino acid is removed and is then added to other carbon chains. In doing so, a brand new amino acid is formed.

❑ Organs that require a large amount of amino acids for protein production are: liver, skeletal muscles, heart, kidneys, and brain.

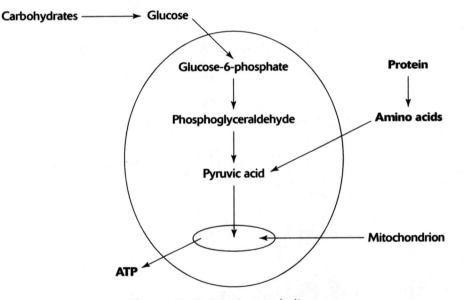

Figure 19-2: Protein metabolism.

Metabolism of Fat

A lot of food ingested consist of fat or is prepared in fat. Fat belongs to a category of molecules called **lipids.** Examples of molecules that fit in the category of lipids are: phospholipids, glycolipids, cholesterol, and fat. Fat molecules are made of a glycerol molecule and three fatty acids. Because it has three fatty acids, it is generally referred to as **triglycerides.** Figure 19-3 illustrates the metabolism of fat, while the following bullet list discusses the metabolism of fat.

Catabolism:

❏ Fat is digested to form glycerol and fatty acids.

❏ Glycerol will enter into the cytosol of the cell and become a component of phosphoglycer-aldehyde (PGAL); hence the "glycer" portion of the name.

❏ Phosphoglyceraldehyde will convert to form pyruvic acid.

❏ Pyruvic acid will enter into the mitochondria.

❏ There are numerous reactions occurring in the mitochondria that lead to the formation of ATP.

❏ The fatty acid component of fat can enter into the mitochondria itself.

❏ Once it has entered the mitochondria, the fatty acids will undergo numerous reactions that lead to the formation of ATP.

Anabolism:

❏ Cells can take the glycerol and fatty acids and recombine them to make a variety of lipids.

❏ Many of such lipids include:

Phospholipids for cell membranes

Glycolipids for the outer layer of cell membranes

Fat for insulation

Cholesterol is not made of glycerol and fatty acids but is, nonetheless, a lipid. Many hormones are cholesterol-based, such as testosterone and estrogen.

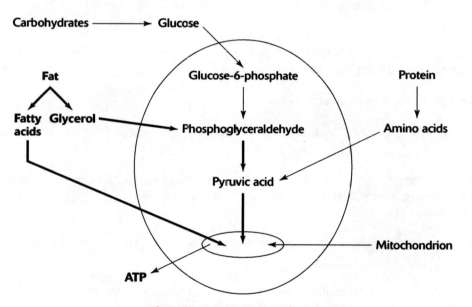

Figure 19-3: Fat metabolism.

Example Problems

1. In reference to the metabolism of carbohydrates, fats, and proteins, there are two molecules that can enter the mitochondria for further reactions. They are:
_____ and _____.

   answer: pyruvic acid; fatty acids

2.　Amino acids are formed from the digestion of _____ and will convert to _____ inside the cell.

answer: protein; pyruvic acid

3.　Due to the activity of _____, one molecule can convert to another molecule during cellular metabolism.

answer: enzymes

4.　Fats are often referred to as triglycerides. This is because they are made of a glycerol molecule and _____.

answer: three fatty acids

5.　Ammonia is a waste product and is produced during the catabolism of _____.

answer: amino acids

Work Problems

1.　Glycolysis is a series of chemical reactions that occur in which part of the cell?

2.　Kreb's reactions are a series of chemical reactions that occur in which cell organelle?

3.　When cells build new organic molecules during metabolism, it is known as _____.

4.　Phosphoglyceraldehyde is made up of a glucose molecule, which is an aldehyde molecule, a phosphate ion, and a _____ molecule.

5.　Glucose is derived from which major food group?

6.　95% if the ATP produced by the cell is produced in the mitochondria and 5% is produced in the _____ of the cell.

7.　Ammonia is produced during metabolism and is quite toxic. To prevent the toxic levels from becoming a problem, the ammonia is converted to _____, which is then excreted in the urine.

8.　When molecules are bonded together to produce a new substance, the process is called _____.

9.　When glucose enters the cell, it needs to change in order to basically stay inside the cell to be converted to other products. Glucose is therefore converted to _____.

10.　The anabolism of amino acids will produce _____.

Worked Solutions

1. **cytosol**

2. **mitochondria**

3. **anabolism**

4. **glycerol**

5. **carbohydrates**

6. **cytosol**

7. **urea**

8. **anabolism**

9. **glucose-6-phosphate**

10. **protein**

Essential and Nonessential Products

The cells of the body have the ability to build (anabolize) a variety of molecules necessary for homeostasis. However, there are some molecules the body needs but the cells cannot make them. Those molecules need to be obtained directly from the diet. The molecules that the body is able to build are called **nonessential** molecules and the ones that the body cannot build (and have to obtain from the diet) are called **essential** molecules.

In reference to amino acids, there are ten essential and ten nonessential amino acids. Table 19-1 lists those amino acids.

Table 19-1 Essential and Nonessential Amino Acids	
Essential Amino Acids	*Nonessential Amino Acids*
Isoleucine	Aspartic acid
Leucine	Glutamic acid
Lysine	Tyrosine
Threonine	Serine
Tryptophan	Cysteine
Phenylalanine	Glycine
Valine	Alanine
Methionine	Asparagine
Arginine*	Glutamine
Histidine*	Proline

*The body makes arginine and histidine but not in adequate amounts (especially in children).

When you study the essential and nonessential amino acids, think in terms like this: It is essential to have a good diet in order to obtain isoleucine, leucine, and so on. In reference to lipids, there are some fatty acid chains the cells can build (nonessential fatty acids) and there are some that the cells cannot build (essential fatty acids). **Linoleic acid** and **linolenic acid** are the most common essential fatty acids; both are necessary for cell membrane fatty acids. Both are necessary for the inflammatory reactions involved in tissue repair.

Current research is also showing that there very well could be essential carbohydrates, too.

Example Problems

1. How do we obtain essential amino acids? _____

    answer: From our diet.

2. How do we obtain nonessential amino acids? _____

    answer: Via metabolism, our body makes the nonessential amino acids.

3. Foods that contains complete proteins are the types of food that has all of the _____ amino acids.

    answer: essential

4. If there were carbohydrates that the body needed but could not make, they would be called _____ carbohydrates.

    answer: essential

5. True or false; The essential amino acids are more important than the nonessential amino acids.

    answer: false. Both types of amino acids are important and necessary for homeostasis. While the nonessential amino acids can be produced by the body, they are no less important than the ones obtained from the diet.

Metabolism of Other Nutritional Products

Proper nutrition requires consuming food that consists of the right amount of carbohydrates, protein, and fat. Also required is obtaining the right amount of essential products. In addition to all of this, one must also obtain the right amount of minerals and vitamins. The body cannot synthesize minerals.

Minerals are in the body in the form of ions:

- ❑ **Calcium ions:** Required for muscle contraction, for nerve impulses, for bone growth, and for blood clotting.

- ❑ **Potassium ions:** Required for normal cardiac activity.

- ❑ **Chloride ions:** Required for normal respiratory activity (chloride shift).

❑ **Magnesium ions:** Required for enzyme formation.

❑ **Sodium ions:** Required for polarized nerve activity.

Vitamins are typically placed in two categories. Some vitamins are fat-soluble and others are water-soluble. The **fat-soluble** vitamins dissolve in lipids and can be stored in the body. Because we are able to store these vitamins, a deficiency of those vitamins is not a typical occurrence. The fat-soluble vitamins are A, D, E, and K. The **water-soluble** vitamins dissolve in water and are therefore easily transported to the kidneys and excreted. Because of this, a deficiency of water-soluble vitamins can be a common occurrence. Water-soluble vitamins are B_1, B_2, B_3, B_6, B_{12}, C, and folic acid.

❑ **Vitamin A:** Necessary for the synthesis of chemicals for vision.

❑ **Vitamins B_1, B_2, and B_3:** All are necessary to complete the chemical reactions in the mitochondria for the formation of ATP.

> B_1 = thiamine
>
> B_2 = riboflavin
>
> B_3 = niacin

❑ **Vitamin B_6:** This is pyridoxine. It is necessary for the metabolism of amino acids and fat.

❑ **Vitamin B_{12}:** This is cobalamin. It is necessary for the red blood cells to absorb iron for the manufacture of hemoglobin.

❑ **Vitamin C:** This is ascorbic acid. It is necessary for collagen formation within tissues.

❑ **Vitamin D:** Necessary for the absorption of calcium ions and phosphorus ions from the small intestine for bone growth and other functions of calcium ions.

❑ **Vitamin E:** Inhibits the breakdown of vitamin A.

❑ **Vitamin K:** Necessary for blood clotting reactions. This can be obtained from the diet but is also produced by bacteria living in the large intestines.

❑ **Folic acid:** Necessary for the metabolism of amino acids and nucleic acids (RNA and DNA).

Vitamins and Metabolism

This section discusses how some of the vitamins are used in the chemical processes of metabolism. Two vitamins of significance are niacin and riboflavin. During the conversion of one organic molecule to another, such as phosphoglyceraldehyde to pyruvic acid, many byproducts are produced. Some of the byproducts are; carbon dioxide (CO_2), hydrogen ions (H^+), water, and amino acids. Much of the carbon dioxide can be exhaled via the respiratory system. The water can always be used by the body's cells. The amino acids produced are the nonessential amino acids. The hydrogen ions become a potential problem. Hydrogen ions have an acidic characteristic. If there are a lot of hydrogen ions generated, the pH could drop. If the pH drops, enzymes necessary for metabolism could cease to function. Metabolism would then stop and ATP would not be produced. Those hydrogen ions need to be buffered. Figure 19-4 shows the byproducts produced and the hydrogen ions being buffered.

Hydrogen ions are produced via the conversion of PGAL to pyruvic acid. The vitamin niacin is converted to NAD (nicotinamide adenine dinucleotide). NAD will bind to the hydrogen ions thereby buffering them. It will then transport the hydrogen ions to specific regions within the mitochondria called the **electron transport system.** The hydrogen ions are then used to manufacture ATP.

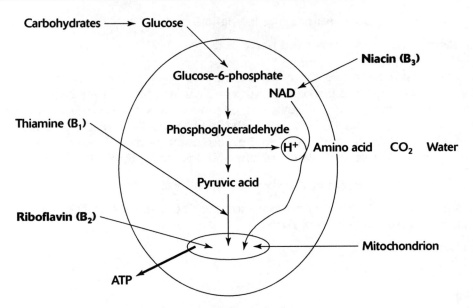

Figure 19-4: Buffering hydrogen ions.

The vitamin riboflavin is converted to FAD (flavin adenine dinucleotide). FAD will bind to the hydrogen ions that are produced within the mitochondria during metabolism, thereby buffering them. It will then transport the hydrogen ions to the electron transport system within the mitochondria. The hydrogen ions are then used in the mitochondria to manufacture ATP. FAD and NAD work in a similar manner to buffer hydrogen atoms. However, the reactions involving NAD will result in the production of more ATP than will reactions involving FAD.

Vitamin B_1 is used to convert pyruvic acid to products that will be used inside the mitochondria. Many nutrition books will state that many vitamins act as buffers during metabolism and some will act as coenzymes for the conversion of organic molecules during metabolism.

Example Problems

1. The B vitamins are water soluble or fat soluble? _____

   answer: water soluble

2. Which two vitamins discussed in this section act as buffers during metabolism?

   answer: niacin and riboflavin

3. What is produced during metabolism that could alter the pH in such a manner to cease the activity of enzymes?

   answer: hydrogen ions

4. Which vitamin is necessary to get calcium ions to leave the small intestine and enter the bloodstream so it can travel to all parts of the body?

   answer: vitamin D

5. Which vitamins are easier for the body to lose; water soluble or fat soluble vitamins?

    answer: water soluble

Cholesterol and Metabolism

Cholesterol is very important in the body. Cholesterol is necessary for the production of cell membranes. It is necessary for the production of some hormones such as testosterone and estrogen. It is also necessary for the formation of vitamin D. The mitochondria of the liver are the organelles that produce cholesterol. Our body makes all the cholesterol it needs. Therefore, excess cholesterol can become a problem. The body deals with this problem with the use of a molecule that has characteristics of a lipid and that of a protein. It is called a **lipoprotein.** This particular lipoprotein is called a **high-density lipoprotein** (HDL). HDL picks up cholesterol and transports it to the liver. Once cholesterol is in the liver, the liver cells will incorporate the cholesterol into bile. Bile will be used by the digestive system and eventually exit the body via waste products. Therefore, by getting rid of bile, the body got rid of excess cholesterol.

However, there is another lipoprotein called LDL **(low-density lipoprotein).** This lipoprotein also binds to cholesterol; however, it seems to drop the cholesterol off in the arteries instead of in the liver. This cholesterol will build up in the arteries thus forming plaque. This is why HDL is considered **good cholesterol** and LDL is considered **bad cholesterol.** In reality, the cholesterol is neither good nor bad; it is how the body handles the cholesterol that is good or bad. Figure 19-5 shows the activity of HDL and LDL. The solid arrows represent the action of HDL and the dotted arrows represent the action of LDL.

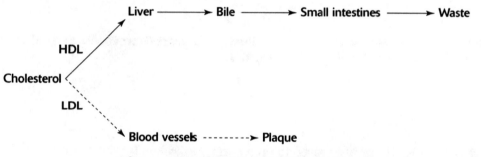

Figure 19-5: The action of HDL and LDL.

Work Problems

1. The bonding of _____ together will result in the formation of a protein molecule.

2. What is the definition of essential amino acids?

3. Which vitamin is produced by bacteria in the large intestine and is used for the production of some blood clotting agents?

4. What is the name of the molecule that ultimately assists in transporting excess cholesterol out of the body?

5. NAD and FAD will bind to hydrogen ions and transport the hydrogen ions to specific regions of the mitochondria, which then metabolize those ions to form ATP. In doing so, NAD and FAD act as _____ to maintain the pH.

6.   Which group of vitamins can be stored in the body?

7.   Why is linoleic acid considered to be an essential fatty acid?

8.   The catabolism of amino acids requires vitamin _____.

9.   There are thousands of different proteins in the body. All of these proteins are made from only _____ different amino acids.

10. Which are we more likely to suffer a deficiency of; the B vitamins or vitamin E?

Worked Solutions

1.   **amino acids**

2.   **Essential amino acids are amino acids that the body cannot make and therefore must be obtained from the diet.**

3.   **vitamin K**

4.   **HDL (high-density lipoprotein)**

5.   **buffers**

6.   **fat-soluble vitamins**

7.   **Linoleic acid cannot be made by the body and must therefore be obtained from the diet, which is the definition of "essential."**

8.   **B$_6$**

9.   **20**

10. **the B vitamins.** These are water-soluble and vitamin E is fat-soluble.

Chapter Problems and Solutions

Problems

1.   Deamination is the process of removing the amine group from an amino acid. This results in the formation of _____.

2.   Urea is formed in the _____ and is excreted by the _____.

3.   Which vitamin is necessary for the production of pigments necessary for vision?

4.   Which vitamin is required for proper absorption of calcium ions from the small intestine into the bloodstream, which can then be used for proper bone growth?

5.   All of the chemical reactions in the body can be summarized with one word. That word is _____.

6. Glycolysis is the name given to the series of chemical reactions that involve glucose in the cytosol of the cell. Therefore, glycolysis ultimately results in the formation of _____.

7. Lipoproteins that transport cholesterol mainly to the liver are called _____.

8. A diet that consists of complete protein means that it contains all the _____ amino acids.

9. Minerals cannot be made by the body therefore they have to come from food source. The body needs chloride ions and sodium ions. Identify the source of sodium and chloride ions.

10. What two major food groups are involved in the production of phosphoglyceraldehyde?

Answers and Solutions

1. **ammonia**

2. **liver; kidneys**

3. **vitamin A**

4. **vitamin D**

5. **metabolism**

6. **pyruvic acid**

7. **high-density lipoproteins (HDL)**

8. **essential**

9. **salt (NaCl) will ionize to form sodium ions and chloride ions.**

10. **Fat digests to form glycerol, which will become a component of PGAL and carbohydrates will digest to form glucose, which will eventually become a component of PGAL.**

Supplemental Chapter Problems

Problems

1. The food we eat comes from plant or animal products. Plant and animal products consist of cells and cells consist of chromosomes, which are made of DNA. Therefore, the body needs to be able to metabolize nucleic acids. What vitamin is necessary to accomplish that task?

2. If niacin and/or riboflavin were lacking in the diet, the pH in the cell that is associated with metabolism would probably (go up or go down) _____.

3. A patient has been found to have poor development of matrix material within the tissues. It was found that this patient has an abnormally low level of which vitamin?

4. Wide-spectrum antibiotics kill the harmful bacteria, which therefore works as an excellent treatment. However, because these antibiotics are considered to be wide-spectrum, they kill even the "good" bacteria. Why would continuous use of wide-spectrum antibiotics cause a decrease in blood clotting activities?

5. A decrease in protein in the diet could also result in the decrease in the formation of which molecule, which therefore would result in a decrease in ATP production?

6. Lipoprotein lipase is an enzyme that breaks down low-density lipoprotein. A drug that blocks lipoprotein lipase would ultimately cause the level of LDL to (go up or go down).

7. Which is better; having higher levels of HDL compared to LDL or having higher levels of LDL compared to HDL?

8. Glycolysis is the name given to the reactions involving the metabolism of glucose in the cytosol of the cell. What would lipolysis be in reference to?

9. A decrease in vitamin D could result in the malfunction of nerves. Explain how this is so.

10. If vitamin A breaks down, it cannot be used for the formation of the pigments involved in vision. What vitamin inhibits the breakdown of vitamin A?

Answers

1. folic acid

2. go down (become more acidic)

3. vitamin C. Vitamin C is necessary to make collagen fibers. Collagen fibers are involved in making a matrix for connective tissue.

4. Wide-spectrum antibiotics will also kill the bacteria living in the large intestine. Those bacteria produce vitamin K, which is necessary for blood clotting.

5. A decrease in protein results in a decrease in amino acids, which therefore results in a decrease in pyruvic acid.

6. LDL levels would go up.

7. Having higher levels of HDL compared to LDL.

8. Lipolysis is the metabolism of fat, which is a lipid, in the cytosol of the cell.

9. Vitamin D is necessary for the absorption of calcium ions into the bloodstream from the small intestines. Without vitamin D, calcium ions cannot enter the bloodstream and cannot be used for the release of neurotransmitters from the presynaptic vesicles in the axons of nerves.

10. vitamin E

Chapter 20
The Urinary System

Waste products are produced during digestion and metabolism. Some of the waste products can leave the body via the respiratory system (such as carbon dioxide), and some can leave via the large intestine in the form of solid waste (feces). Some of the waste products leave via the urinary system in the form of liquid waste. The primary function of the urinary system is to get rid of liquid waste, but the kidneys have more functions than just waste removal.

The Functions and Structures of the Urinary System

The kidneys perform numerous functions. Most people are familiar with the fact that the kidneys are involved in the removal of liquid waste. Consider the following functions of the urinary system:

❑ Gets rid of liquid waste.

❑ Regulates blood volume and therefore blood pressure via the action of ADH and ANP and Aldosterone.

❑ Regulates blood pH by removing hydrogen ions

❑ Causes the formation of erythrocytes via the action of EPO.

❑ Prevents dehydration.

Waste products will enter into the circulatory system and will travel to the kidneys via the renal artery. Once inside the kidneys, the waste products will enter into numerous tubules called **nephrons.** The nephrons will "process" the waste and will send the waste to the tubes that exit the kidneys called the left and right **ureters.** Each ureter will transport waste products to the **urinary bladder.** Liquid waste can be stored in the urinary bladder until the appropriate time to void. Urine will leave the urinary bladder via the single **urethra.** Figure 20-1 shows the structures of the urinary system.

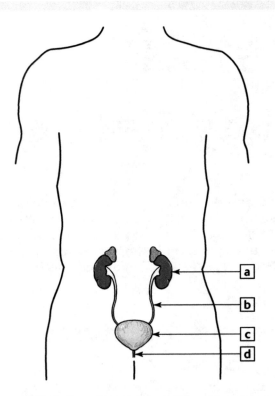

Figure 20-1: Overall structures of the urinary system.

Example Problems
Use the information in this section to answer the following questions.

1. What is the name for structure a?

   answer: kidney (left kidney)

2. What is the name for structure b?

   answer: ureter (left ureter)

3. What is the name for structure c?

   answer: urinary bladder

4. What is the name for structure d?

   answer: urethra

The Internal Structures of the Kidney

The main functioning units of the kidney are the numerous tubules called **nephrons.** Waste products enter into the kidney via the renal artery. From there, the waste enters into several smaller blood vessels ultimately leading to small capillaries called the **glomerular capillaries.** Waste products are forced (due to blood pressure) out of the capillaries and into the first part of the nephron called the **glomerular capsule** (formerly called Bowman's capsule). Waste products will then enter into a coiled tube of the nephron called the **proximal convoluted tubule** (PCT). The waste products will continue through the nephron into the **nephron loop** (formerly called the loop of Henle). Waste products will continue through to another coiled tube called the **distal convoluted tubule** (DCT). From there, the waste will enter into the **collecting duct,** which collects waste from the DCT of several other nephrons. While passing through all the regions of the nephron, the waste is being processed. Water is being removed, solutes are being removed and added, and so on. From the collecting ducts of the kidneys, waste will eventually leave the kidneys and enter into the ureters.

The collecting duct passes through the renal pyramid region. All of the collecting ducts passing through a common **renal pyramid** will dump waste into a **minor calyx.** Waste from several minor calyces will "dump" into a common **major calyx.** Several major calyces will dump the waste collectively into the **renal pelvis** region. The renal pelvis leads to the ureter, which in turn leads to the urinary bladder. Figure 20-2 illustrates the internal parts of a kidney, and Figure 20-3 illustrates the parts of a nephron and its associated structures.

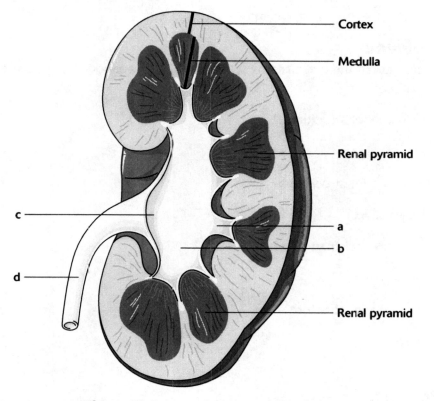

Figure 20-2: Internal structures of the kidney.

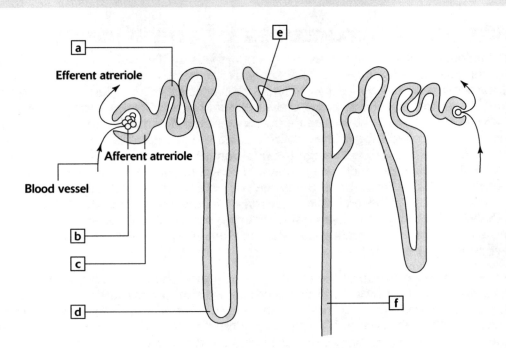

Figure 20-3: Parts of a nephron.

Example Problems

Use the information about the internal structures of the kidney to identify the structures in Figure 20-2.

1. What is the name for structure a?

    answer: minor calyx

2. What is the name for structure b?

    answer: major calyx

3. What is the name for structure c?

    answer: renal pelvis

4. What is the name for structure d?

    answer: ureter

Example Problems

Use the information about the nephron to identify the structures in Figure 20-3.

1. What is the name for structure a?

    answer: proximal convoluted tubule

2. What is the name for structure b?

   answer: glomerular capillaries

3. What is the name for structure c?

   answer: glomerular capsule

4. What is the name for structure d?

   answer: nephron loop

5. What is the name for structure e?

   answer: distal convoluted tubule

6. What is the name for structure f?

   answer: collecting duct or collecting tubule

Work Problems

1. Urine is transported to the urinary bladder via what tubes?

2. The renal artery and vein enter and exit the kidney at a depression called the hilus. What tube also exits in that same area?

3. The collecting tubules will transport waste from the various nephrons to the minor calyx by passing through which structure of the kidney?

4. The first part of the nephron is called the _____.

5. The main functional unit of the kidney is called the _____.

6. What is the name of the capillaries located between the afferent arteriole and the efferent arteriole?

7. What is the name of the portion of the nephron that connects to the collecting duct?

8. In order to prevent all the water from leaving the body (and therefore dehydration); a lot of the water is put back into the bloodstream by leaving the nephron and reentering the bloodstream. This action is due to the hormones such as aldosterone and _____.

9. Urine is transported out of the urinary bladder and therefore out of the body via which tube?

10. Which kidney sits higher in the body?

Worked Solutions

1. **ureters**

2. **ureters**

3. **renal pyramids**

4. **glomerular capsule**

5. **nephron**

6. **glomerular capillaries**

7. **distal convoluted tubule (DCT)**

8. **ADH (antidiuretic hormone)**

9. **urethra**

10. **left kidney sits higher than the right kidney**

Blood Vessels Associated with the Nephron

A major job of the kidneys, but yet just one of the many jobs, is to get rid of liquid waste. Liquid waste enters the kidneys via the renal artery. From there, the waste travels through a series of blood vessels leading to the glomerular capillaries.

The blood vessel previous to the glomerular capillaries is the **afferent arteriole.** There is a blood vessel that leaves the glomerular capillaries, which is called the **efferent arteriole.** The efferent arteriole leads to the **vasa recta,** which is a series of blood vessels that loop around the entire nephron (for simplicity, the vasa recta is not shown wrapping around the nephron). The vasa recta leads ultimately to the renal vein. The renal vein enters into the inferior vena cava, which takes blood back to the right atrium of the heart. Figure 20-4 illustrates the blood vessels associated with the nephron.

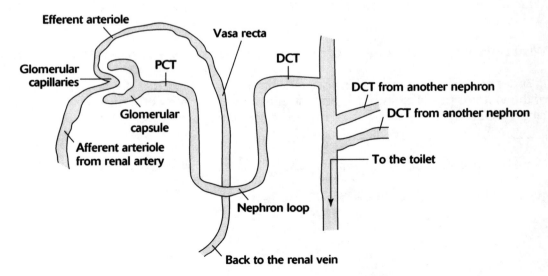

Figure 20-4: Blood vessels and the nephron.

Function of the Nephron

Figures 20-5, 20-6, and 20-7 illustrate how a nephron works. A major function of the nephron is to get rid of liquid waste. Figure 20-5 illustrates how waste products eventually enter into the glomerular capillaries. The waste material will then pass through the various regions of the nephron and ultimately pass through the collecting tubule, which will send the urine to the ureter. The ureter will then send the urine to the urinary bladder. The waste material will leave the urinary bladder by exiting via the urethra.

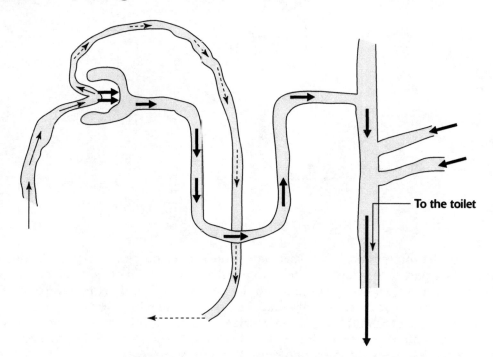

Figure 20-5: The nephron and waste material.

Waste material is produced by the various cells of the body. The blood will pick up the waste material. The waste material will eventually arrive at the renal artery. The renal artery transports the waste to the internal regions of the kidney where it eventually forms the afferent arteriole. Due to blood pressure, the waste products are literally forced out of the glomerular capillaries and into the glomerular capsule. To help create the blood pressure needed to force waste into the glomerular capsule, the efferent arteriole has a smaller diameter than the afferent arteriole. The dark arrows represent the fact that the majority of the waste material will pass through the nephron to eventually exit the body. Notice that the efferent arteriole has a dotted line. This represents the fact that not all of the waste that enters the glomerular capillaries is forced into glomerular capsule. Some of the waste stays in the circulatory system. This is a small amount, however. This small amount of waste material flows in the efferent arteriole and then into the vasa recta. The vasa recta eventually leads to the renal vein and then to the inferior vena cava. Therefore, the renal artery transports a lot of waste into the kidney and the renal vein transports just a small amount of waste out of the kidney.

Figure 20-6 illustrates how water passes through the nephron. One function of the kidneys is to prevent dehydration. The plasma of blood consists of about 92% water. There is a lot of water entering into the nephron. If all this water were to leave the kidneys, we would be very dehydrated. There is approximately 50 gallons of water entering into the kidneys every day! If we lost 50 gallons of water per day, we would be severely dehydrated. The kidneys are very efficient at preventing dehydration.

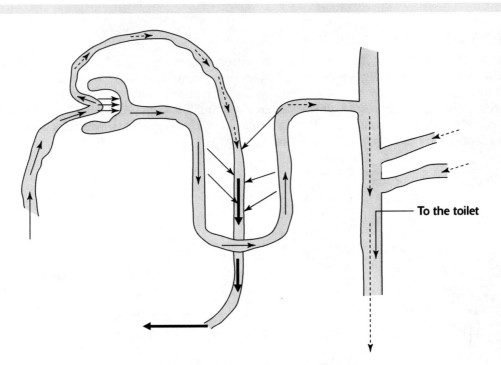

To the toilet

Figure 20-6: The nephron and water.

Water enters into the afferent arteriole. This water enters into the glomerular capillaries and is forced into the first part of the nephron (the glomerular capsule). To prevent all of this water from leaving the kidneys, some of it enters into the vasa recta. The dark, bold arrows in the vasa recta represent the fact that there is a large amount of water leaving the nephron and entering into the vasa recta. In fact, about 99% of the water enters into the vasa recta and therefore stays in the body. Only about 1% of the water actually exits the body. The antidiuretic hormone (ADH) causes the water to leave the nephron and enter into the vasa recta.

Figure 20-7 illustrates the fact that blood should not appear in the urine, because blood cannot enter the glomerular capsule under normal conditions. If blood is found in the urine, this means that blood has either entered into the nephron, the ureters, or the urethra. If blood enters the nephron, it is typically due to a bacterial infection that has damaged the lining of the glomerular capsule.

In Figure 20-7, notice that the blood does not enter the glomerular capsule. Therefore, all the blood that has entered the glomerular capillaries will eventually enter into the vasa recta and then into the renal vein. From the renal vein, the blood will enter the inferior vena cava and return to the heart.

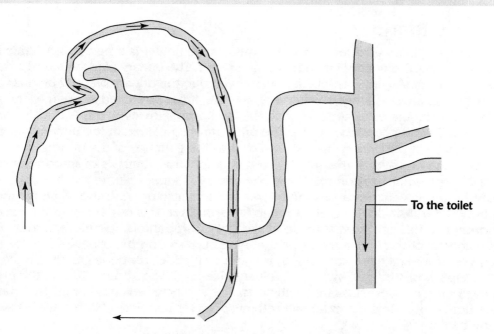

Figure 20-7: The nephron and blood.

Example Problems

Use the information in Figures 20-5, 20-6, and 20-7 to answer the following questions.

1. In order to prevent dehydration, water in the nephron will leave the nephron and enter into the _____.

   answer: vasa recta

2. Which contains "cleaner" blood; the afferent arteriole or the efferent arteriole?

   answer: efferent arteriole; the majority of the waste is forced into the glomerular capsule before entering the efferent arteriole.

3. Which contains a greater amount of waste; the afferent arteriole or the vasa recta?

   answer: afferent arteriole

4. If there were 100 erythrocytes in the afferent arteriole, under normal conditions, how many erythrocytes would there be in the efferent arteriole?

   answer: 100. Erythrocytes should not leave the circulatory system

5. The vasa recta eventually leads to the renal vein. The renal vein leads to the

   _____.

   answer: inferior vena cava (IVC)

The Urinary Bladder

The function of the urinary bladder is to store liquid waste until it is time to void. Urine leaving the kidneys will enter the urinary bladder via the ureters. The urinary bladder fills at a rate of about 80 mL per hour. As urine enters the urinary bladder, the urinary bladder begins to stretch. When the urinary bladder reaches a volume of 200 mL, it has stretched enough to trigger a signal giving a person the initial urge to go to the bathroom. However, this is not a very strong urge. The urinary bladder continues to fill. The urine stays contained in the urinary bladder because at the base of the urinary bladder, which is at the entrance of the urethra, there is a sphincter called the **internal urethral sphincter.** This sphincter consists of smooth muscle and is therefore under involuntary control. When the urinary bladder volume reaches about 500 mL, the pressure from this volume causes the internal sphincter to open. Urine will now enter the urethra. About 2 cm distal in the urethra is another sphincter. This one is called the **external urethral sphincter.** This sphincter is made of skeletal muscle and is, therefore, under voluntary control. The control of this sphincter keeps the urine from exiting the urethra until the appropriate time arrives. If we still have not voided, the urinary bladder continues to fill and stretch. The urge to void is getting stronger. When the urinary bladder fills to about 800 mL, the pressure is too great even for the external urethral sphincter. This sphincter will now open whether it is the appropriate time or not. It is the external urethral sphincter that little children learn to control during potty training.

Work Problems

1. When approximately _____ mL of urine enters the urinary bladder, the urinary bladder will stretch enough to give the initial urge to go to the bathroom.

2. We do not have voluntary control over which urethral sphincter, even as adults?

3. As infants, we do not have voluntary control over either of the urethral sphincters. However, as we age, we eventually learn to control which urethral sphincter?

4. If 50 gallons of filtrate enter the kidneys per day, how much is actually lost to the toilet?

5. A lack of ADH would cause the kidneys to put excess water into the _____ (vasa recta or collecting tubule).

6. What part of the nephron is involved in filtration?

7. Material too large to enter into the glomerular capsule will enter into which blood vessel?

8. The collecting ducts pass through which part of the kidney?

9. Hematuria is a term that refers to _____.

10. The main functioning unit of the kidney is the _____.

Worked Solutions

1. **200**

2. **internal urethral sphincter**

3. **external urethral sphincter**

4. **$50 \times 0.01 = 0.5$ gallons**

5. **collecting tubule**

6. **glomerular capsule**

7. **efferent arteriole**

8. **renal pyramid**

9. **blood in the urine**

10. **nephron**

Other Functions of the Kidneys

The main function of the kidney is the removal of waste. This section discusses some of the other functions of the kidneys. Table 20-1 lists a few select functions of the kidneys, and Table 20-2 discusses some of the responses performed by the kidneys.

Table 20-1 Additional Functions of the Kidneys	
Function	**Brief Discussion**
Balancing body fluids	Water moves back and forth across the cell membranes. If the kidneys were to excrete excess water, this would result in less fluid in the blood. In order to maintain fluid level in the blood, the cells would lose water to the blood. The cells would then dehydrate.
Electrolyte balance	Electrolytes are substances that produce ions such as when NaCl ionizes to form Na^+ and Cl^-. These ions are often referred to as minerals. If the kidneys were to lose too many minerals (ions) this would cause the bloodstream to be hypotonic. This osmotic situation would cause the cells to lose water to the bloodstream thus resulting in dehydration.
Acid-base balance	Due to cellular metabolism, the body is always producing CO_2. This CO_2 always binds to water thus producing carbonic acid, which will ionize to form hydrogen ions, which have acidic characteristics. The kidneys will excrete some of the hydrogen ions to help balance the pH. Therefore, urine pH is typically acidic.
Balances blood pressure	The loss of water via the kidneys will lower the blood pressure. Therefore, the retention of water via the kidneys will raise the blood pressure. The balance between water loss and retention will result in a balanced blood pressure value.
Formation of erythrocytes	The juxtamedullary cells of the kidneys release erythropoietin upon detection of low oxygen content in the blood. EPO will target the bone marrow and begin the process of making more erythrocytes in an effort to raise the oxygen level arriving at the kidney cells.

Table 20-2 Kidney Responses	
Response Action	**Brief Discussion**
Erythropoiesis	The juxtamedullary cells of the kidneys respond to a lack of oxygen by releasing erythropoietin. EPO eventually causes the bone marrow cells to begin producing erythrocytes.
Renin/angiotensin system	In response to low blood pressure, the juxtamedullary cells also release rennin. Renin activates angiotensin II. Angiotensin II causes vasoconstriction, which increases blood pressure. Angiotensin II also stimulates the release of ADH and aldosterone. ADH and aldosterone eventually result in the retention of water, and thus raises blood pressure.
Aldosterone	Aldosterone causes the release of sodium ions from the nephron to the vasa recta. The vasa recta increases in solute, which creates a hypertonic region. Water will then leave the nephron and enter into the vasa recta, thereby raising blood volume and ultimately blood pressure.
Antidiuretic hormone	ADH is released from the hypothalamus due to a fall in blood pressure. ADH causes the nephron to put water into the vasa recta. This increases blood volume and ultimately blood pressure.
Atrial natriuretic peptide	ANP is released from atrial cells of the heart in response to an increase in blood volume. ANP targets the nephron and causes the nephron to retain sodium ions. This causes the nephron to become hypertonic to the vasa recta. Water will leave the vasa recta and enter the nephron. As water leaves the vasa recta, volume is lost and, therefore, blood pressure begins to drop.

Example Problems

1. What cells of the kidney respond to low blood pressure? _____

    answer: juxtamedullary cells

2. The antidiuretic hormone is released when the cells of the hypothalamus begin to lose water. What else causes the release of the antidiuretic hormone?

    answer: angiotensin II

3. When the juxtamedullary cells of the kidneys detect adequate amounts of oxygen, the release of erythropoietin will (increase or decrease).

    answer: decrease. EPO is released only when the oxygen level decreases.

4. Aldosterone will cause the vasa recta to become _____ and the nephron to become _____. (Use osmotic terms to fill in the blanks).

    answer: hypertonic; hypotonic

5. Urine typically has a pH around 6.0 because the kidneys urinate _____ ions.

    answer: hydrogen

Chapter Problems and Solutions

Problems

1. The urinary bladder will fill approximately 80 mL per hour. Approximately how many hours would pass before the original urge to go to the bathroom occurs?

2. Atrial natriuretic peptide is released when blood pressure is (high or low).

3. Antidiuretic hormone is released when blood pressure is (high or low).

4. As urine flows out of the kidneys and is transported to the urinary bladder, it travels through what tubes?

5. The depression where the renal artery enters the kidneys and the renal vein and ureters exit the kidneys is called the _____.

6. Material in the glomerular capsule will enter the _____ next.

7. If 200 mL of water enters the nephron, how many mL will normally be put back into the vasa recta?

8. A rise in carbon dioxide will cause the blood pH to (rise or fall) _____.

9. What is the name of the capillaries that surround the nephron?

10. Put the following terms in proper sequence regarding the flow of urine through the kidney and to the toilet: minor calyx, collecting duct, ureter, major calyx, urethra, urinary bladder, renal pelvis

Answers and Solutions

1. **2.5 hours.** The urinary bladder fills at a rate of 80 mL per hour and it takes approximately 200 mL to give the original urge; therefore, 200 mL divided by 80 mL per hour = 2.5 hours.

2. **high.** ANP ultimately causes the kidneys to put water into the circulatory system.

3. **low.** ADH ultimately causes the kidneys to retain water, thereby bringing blood pressure down to normal.

4. **ureters**

5. **hilus**

6. **proximal convoluted tubule (PCT)**

7. **99% is reabsorbed into the vasa recta.** Therefore, 99% of 200 mL = 198 mL

8. **fall (becomes too acidic)**

9. **vasa recta**

10. **collecting duct, minor calyx, major calyx, renal pelvis, ureter, urinary bladder, urethra**

Supplemental Chapter Problems

Problems

1. Red blood cells in the glomerular capillaries will enter the _____ next.

2. A drastic drop in blood pressure at the glomerular capillary region will (increase or decrease) the amount of waste entering the glomerular capsule.

3. When a person holds their breath, will the blood pH go up or down?

4. As we age, we could lose control over which urethral sphincter?

5. Having low blood pressure could affect the efficiency of the kidneys because the glomerular capillaries may put (excess or not enough) waste material into the glomerular capsule.

6. Having chronic high blood pressure could affect the efficiency of the kidneys because the glomerular capillaries may put (excess or not enough) material, into the glomerular capsule.

7. Cystitis is an inflammation of which of the urinary structures?

8. What is the medical term that refers to painful urination?

9. Having excess glucose in the urine is called _____.

10. An obstruction in the glomerular capillaries will affect the flow of blood into the _____.

Answers

1. efferent arteriole

2. decrease

3. down. Holding the breath will increase carbon dioxide in the blood (CO_2 plus water produces H^+).

4. external

5. not enough

6. excess

7. urinary bladder

8. dysuria

9. glycosuria

10. efferent arteriole

Chapter 21
The Reproductive System

M ost systems of the body require other systems for homeostasis. For example, in order for the respiratory system to function properly, a portion of the muscular system is required. The muscles are needed to inhale and exhale. The urinary system relies on the circulatory system. The circulatory system transports waste to the urinary system. There aren't any systems of the body that require the assistance of the reproductive system; however, the reproductive system requires the assistance of all other systems of the body. The reproductive system needs nutrients from the digestive system, the cells of the reproductive system need oxygen from the respiratory system, and so on. This chapter discusses how the reproductive system is involved in perpetuating the species.

The Male Reproductive System and Sperm Cells

Follicle stimulating hormone (from the adenohypophysis—anterior pituitary gland) targets the **seminiferous tubules** inside the testes and initiates sperm production. Sperm cells leave the seminiferous tubules and travel to the **epididymis.** From the epididymis, the sperm cells enter the **ductus deferens** (also known as the vas deferens). The ductus deferens is a long tube that transports the sperms cells to the **ejaculatory duct** that enters into the **prostate gland.** The ejaculatory duct connects to the **penile urethra** that will transport the sperm out of the penis. Figure 21-1 illustrates the male reproductive structures discussed in this paragraph.

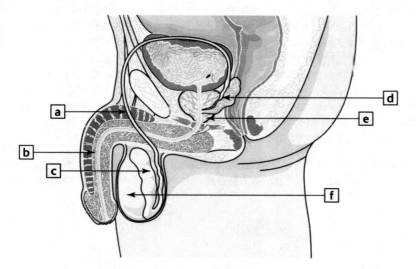

Figure 21-1: Male reproductive structures.

Example Problems

Use the information in the preceding paragraph to answer the following questions regarding Figure 21-1.

1. What is the name for structure a?

   answer: ductus deferens

2. What is the name for structure b?

   answer: penile urethra

3. What is the name for structure c?

   answer: epididymis

4. What is the name for structure d?

   answer: ejaculatory duct

5. What is the name for structure e?

   answer: prostate gland

6. What is the name for structure f?

   answer: testis with the seminiferous tubules inside

The Three Glands of the Male Reproductive System

As the sperm cells travel through the male's body, they encounter three major glands. As the sperms cells are traveling in the ductus deferens, they swim past the first gland called the **seminal vesicle.** The seminal vesicles provide semen that contains fructose that is necessary for sperm survival. The semen (with fructose) from the seminal vesicle will enter the pathway of sperm cells by entering into the ejaculatory duct. The sperm cells continue to enter into the penile urethra and pass through the second gland called the **prostate gland.** The prostate gland surrounds the urethra as it leaves the urinary bladder and enters into the penis (penile urethra). The prostate gland also provides semen but this semen is slightly acidic. The sperm cells continue to pass through the penile urethra and will pass by the third gland called the **bulbourethral gland.** The bulbourethral gland ejects its slightly alkaline semen into the pathway of the sperm cells as they are swimming through the penile urethra.

The male will ejaculate about 200 million to 600 million sperm cells per ejaculate. The male also ejaculates about 2–5 mL of semen. Figure 21-2 illustrates the three glands of the male reproductive system.

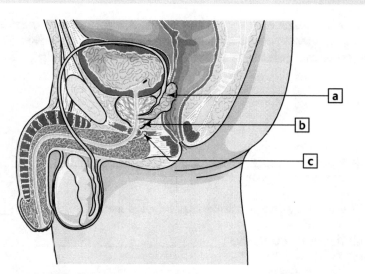

Figure 21-2: The accessory glands of reproduction.

Example Problems

Use the information in the previous paragraph to answer the following questions regarding Figure 21-2.

1. What is the name for structure a?

    answer: seminal vesicle gland

2. What is the name for structure b?

    answer: prostate gland

3. What is the name for structure c?

    answer: bulbourethral gland

Structure of Sperm Cells

A sperm cell is made of the head and a tail called the **flagellum.** The flagellum moves in such a manner to propel the sperm cells through the male's reproductive system. The head of the sperm cells consists of a cap called the **acrosomal cap.** This cap consists of an enzyme called **hyaluronidase.** This enzyme is necessary to decompose the cellular barrier of the egg to allow for penetration of the sperm into the egg. Once the sperm has entered the egg, the egg is fertilized.

Hormones and Nerves Associated with the Male Reproductive System

The **follicle stimulating hormone** (from the adenohypophysis—anterior pituitary gland) will target the seminiferous tubules to begin the formation of sperm cells. The **luteinizing hormone** (also from the adenohypophysis) will target specific cells within the testes called the interstitial cells. This hormone will cause the interstitial cells to release testosterone. Testosterone will target other cells of the body to create the secondary sex characteristics of males.

A **parasympathetic nerve** causes erection of the penis. A **sympathetic nerve** causes ejaculation of semen from the various glands and the ejaculation of sperm and semen from the penis. The sympathetic nerves also cause peristalsis of the ductus deferens to help propel the sperm cells through the tube.

Work Problems

1. Due to the action of the _____, sperm cells will begin to form in the seminiferous tubules.

2. The sperm cells will leave the seminiferous tubules and enter into the _____.

3. Sperm cells will travel through the _____ to enter into the ejaculatory duct.

4. Semen from the _____ will enter into the ejaculatory duct and provide the sperm cells with fructose.

5. The urethra from the urinary bladder is joined by the ejaculatory duct and will pass through the _____ gland before entering the penis.

6. As the sperm cells exit the prostate region, they swim past a small gland called the _____ gland and will enter into the penile urethra.

7. The _____ nerve will cause an erection and the _____ nerve will cause ejaculation.

8. The _____ hormone will cause the development of sperm cells and the _____ hormone will ultimately cause the release of testosterone.

9. Hyaluronidase is an enzyme found in the head region of the sperm cell called the _____.

10. What is the specific part of the testes that produces sperm cells?

Worked Solutions

1. **follicle stimulating hormone**

2. **epididymis**

3. **ductus deferens**

4. **seminal vesicle**

5. **prostate**

6.   **bulbourethral**

7.   **parasympathetic; sympathetic**

8.   **follicle stimulating; luteinizing**

9.   **acrosomal cap**

10.   **seminiferous tubules**

The Female Reproductive System and Egg Cells

At the time of birth, the female's ovaries already consist of immature egg cells. At puberty, the follicle stimulating hormone (from the adenohypophysis) will target the ovaries and cause the immature eggs to mature. There will then be a surge of luteinizing hormone (from the adeno-hypophysis) that will cause a mature egg to ovulate. This egg will leave the follicle it is in and enter into the uterine tube. The egg does not have any means of locomotion like the sperm cells have. Therefore, the egg relies on the cilia inside the uterine tube to move it along the length of the tube.

If the egg is not fertilized, it will begin to decompose about two days after ovulation. If it is fertilized, the egg will continue to travel down the uterine tube and will bury itself in the endometrial lining of the uterus.

The uterus itself is made of an inner lining called the **endometrium** and a muscular portion called the **myometrium.** The egg buries itself in the endometrial lining and continues to develop. The myometrium is made of smooth muscle and causes uterine contractions during childbirth.

Sperm Cells Fertilizing the Egg

Sperm cells will exit the male's penis and pass through the cervix of the uterus. The sperm cells will swim through the body of the uterus and will enter into the uterine tubes. They will swim the length of the uterine tubes until they come in contact with an egg. The sperm cells will bind to the outer cells of the egg.

As the egg is ovulated and, therefore, leaving the ovarian follicle, it becomes surrounded by a cluster of cells called the **corona radiata.** These cells provide protection for the egg. Therefore, the sperm cells have to break down cells of the corona radiata in order to penetrate the egg and thereby fertilizing it.

Each sperm cell binds to the egg and begins to release its stored hyaluronidase. The hyaluronidase will begin to decompose the protective layer of the egg. Once the protective layer decomposes, a sperm cell can enter and fertilize the egg.

In order for successful fertilization to occur, the sperm cells need to "find" the egg in the distal two-thirds of the uterine tube. Figure 21-3 shows a frontal view of the uterus and the uterine tubes. The large dot in the right uterine tube represents the ovulated egg. The dotted arrows represent the pathway of sperm cells swimming to fertilize the egg.

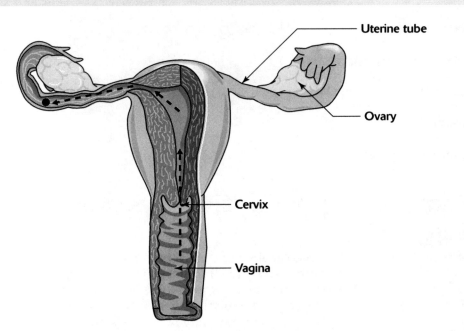

Figure 21-3: Frontal view of the uterus.

The Fertilized Egg

The fertilized egg is called a **zygote.** The zygote will continue its journey in the uterine tube on its way to the inside lining of the uterus called the endometrium. The endometrium has thickened in preparation of the zygote. The zygote will eventually embed itself into the endometrium and continue to develop. Figure 21-4 shows the travel of the zygote and its eventual embedding into the endometrial lining.

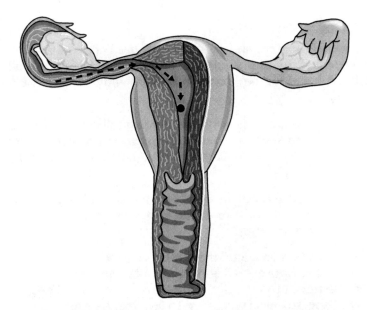

Figure 21-4: Embedding into the endometrial lining.

Example Problems

1. Sperm cells do not fertilize the egg in the body of the uterus. They fertilize the egg in the
   _____.

   answer: distal 2/3 of the uterine tube

2. The inside lining of the uterus is called the _____.

   answer: endometrium

3. The muscular portion of the uterus that is involved in labor contractions is the
   _____.

   answer: myometrium

4. What is the name of the enzyme necessary to break down the egg barrier?

   answer: hyaluronidase

5. What is the name of the cells that provide protection and therefore act as a barrier for the egg cells?

   answer: corona radiata

Hormones Associated with the Female Reproductive System

At puberty, the **follicle stimulating hormone** (from the adenohypophysis), also called **FSH,** will target the ovaries and cause the immature eggs to mature. The **luteinizing hormone** (from the adenohypophysis) will cause a mature egg to ovulate. The luteinizing hormone causes the follicle that has the egg to rupture. This ruptured follicle is called a **corpus luteum.** The corpus luteum now releases progesterone and estrogen. Progesterone will target the endometrial cells of the uterus. This hormone will cause the endometrial lining to thicken in preparation for a fertilized egg.

The zygote itself produces a hormone. This hormone is called **human chorionic gonadotropin.** This hormone will target the corpus luteum to ensure that it will continue to produce progesterone. High levels of progesterone are necessary to maintain a thick endometrial lining for the survival of the zygote. The buried zygote will eventually produce a placenta. The placenta itself will produce progesterone. This progesterone also targets the endometrial lining. Extremely high levels of progesterone also target the adenohypophysis and has the function to "turn off" the release of follicle stimulating hormone. This is why (during pregnancy) a woman cannot produce and release a second egg.

Work Problems

1. What hormone is involved in causing the development of a mature egg?

2. What hormone is involved in ovulation?

3. When an egg becomes fertilized, it is called a _____.

4. After the egg is fertilized, it will eventually travel to and embed itself into the
   _____.

5. What structure of the ovaries releases progesterone? _____

6. Progesterone targets the _____ and extremely high levels will target the _____ .

7. How does the fertilized egg travel to the endometrial lining?

8. A ruptured follicle is called a _____ and it releases _____ .

9. What hormone ensures that the corpus luteum will continue to maintain high levels of progesterone?

10. Successful fertilization of the egg occurs in which female reproductive structure?

Worked Solutions

1. **follicle stimulating hormone**

2. **luteinizing hormone**

3. **zygote**

4. **endometrial lining**

5. **corpus luteum**

6. **endometrial lining; adenohypophysis**

7. **The fertilized egg travels via the action of cilia lining the uterine tubes.**

8. **corpus luteum; progesterone**

9. **human chorionic gonadotropin**

10. **uterine tube (distal ⅔ of the uterine tube)**

General Reproductive System Information

Table 21-1 lists some select information regarding the male and female reproductive system. The information is briefly discussed as a summary of the reproductive systems.

Table 21-1 Select Information Regarding the Reproductive System	
Select Topic	**Discussion**
Pregnancy test	The presence of HCG (human chorionic gonadotropin) is an indication of pregnancy. If HCG is present, that means a zygote is present. If a zygote is present, the person is indeed pregnant.
Meiosis	This is the special type of cell reproduction that is involved in the development of sperm cells and egg cells. These cells are called **gametes.** Only gametes are involved with meiosis.

Select Topic	Discussion
Spermatogenesis	This is the process of meiosis that develops sperm cells.
Oogenesis	This is the process of meiosis that develops egg cells.
Vasectomy	This is the process of cutting the ductus deferens tubes (vas deferens). Sperm is still produced in the seminiferous tubules but cannot travel very far in the ductus deferens. Sperm cells therefore cannot come in contact with the egg. The male will still be able to ejaculate semen however.
Tubal ligation	This is the process of cutting the uterine tubes. The female will still produce and ovulate eggs. The eggs cannot travel very far in the uterine tube and therefore will not come in contact with sperm cells.
Menstruation	A decrease in progesterone will ultimately result in the loss of the thickened portion of the endometrium. Progesterone will begin to decrease when the corpus luteum ceases progesterone production. The corpus luteum will stop producing progesterone after about 10 days if the person does not develop a zygote.
Menopause	This is the time period when the adenohypophysis no longer produces the follicle stimulating hormone. Without FSH, an egg will not develop.

Chapter Problems and Solutions

Problems

1. Mitosis is the type of cell reproduction that involves most cells of the body. _____ is the type of cell reproduction that involves the formation of the gametes.

2. True or false: Sperm cells enter into the seminal vesicle gland and then enter into the ejaculatory duct. _____

3. Spermatogenesis occurs in the _____.

4. Oogenesis occurs in the _____.

5. What hormone causes the release of testosterone?

6. After a male has a vasectomy, he can still ejaculate material. What is he ejaculating?

7. The onset of menstruation is due to a decrease in what hormone?

8. The onset of menopause is due to a decline in what hormone?

9. During pregnancy, the woman typically does not produce another egg. This is due to the excessively high production of _____, which will inhibit the production of the follicle stimulating hormone.

10. What is the name of the gland that actually surrounds the urethra in the male reproductive system?

Answers and Solutions

1. **Meiosis**

2. **false.** Sperm cells swim past the gland; they do not enter it.

3. **seminiferous tubules**

4. **ovaries**

5. **luteinizing hormone**

6. **He is ejaculating semen, not sperm cells.**

7. **progesterone**

8. **follicle stimulating hormone**

9. **progesterone**

10. **prostate gland**

Supplemental Chapter Problems

Problems

1. If the male produces 5 mL of semen and 500 million sperm cells, how many sperm cells are there per mL?

2. Which is pleural; testes or testis?

3. In the world of anatomy, anatomists are trying to get rid of anatomical structures that have a person's name. Therefore, the Achilles' heel is now called the calcaneal tendon. In reference to the reproductive system, what is the correct anatomical name for the formerly named Fallopian tubes?

4. True or false: After a vasectomy, the male will not be able to develop an erection.

5. True or false: Tubal ligation prevents egg production and ovulation.

6. The male gamete is the _____.

7. What organs are involved in the production of semen?

8. The nucleus of human cells consists of 46 chromosomes. The process of meiosis will make it so the combination of a sperm cell with an egg cell will create a total of 46 chromosomes. Therefore, each sperm cell and each egg cell must have _____ chromosomes.

9. If a patient experiences prostate swelling, this could cause a constriction of the
   _____, which makes urination and ejaculation difficult.

10. Due to parasympathetic action, the blood vessels in the penis will (dilate or constrict) thus
    resulting in an erection.

Answers

1. 100 million per mL

2. testes

3. uterine tubes

4. false. A vasectomy is the cutting of the ductus deferens, which prevents sperm cell travel.
   An erection is developed via parasympathetic nerve activity.

5. false. Tubal ligation is the cutting of the uterine tubes. Egg production and ovulation are
   the result of hormone activity.

6. sperm cell

7. seminal vesicle, prostate gland, and bulbourethral gland

8. 23

9. urethra

10. dilate

Customized Full-Length Exam

1. How are the hands positioned in the anatomical position?

    Answer: The hands are at the side of the body with the palms facing anterior.

If you answered **correctly,** go to question 4.
If you answered **incorrectly,** go to question 2.

2. When you discuss a patient, you refer to thumbs as _____ structures compared to the position of the little finger.

    Answer: lateral

If you answered **correctly,** go to question 4.
If you answered **incorrectly,** go to question 3.

3. When you discuss a patient, you refer to the hallux as a _____ structure compared to the position of the little toe.

    Answer: medial

If you answered **correctly,** go to question 4.
If you answered **incorrectly,** review "Directional Terms" on pages 37–38.

4. What is the anatomical term that refers to the mouth?

    Answer: Oris

If you answered **correctly,** go to question 6.
If you answered **incorrectly,** go to question 5.

5. What is the name of the first part inferior to the oris?

    Answer: Mentis

If you answered **correctly,** go to question 6.
If you answered **incorrectly,** review "Superficial Landmarks" on page 42.

6. The armpit area is called the _____.

    Answer: axilla

If you answered **correctly,** go to question 8.
If you answered **incorrectly,** go to question 7.

7. The wrist area is called the _____ and the ankle area is called the _____.

    Answer: carpal; tarsal

If you answered **correctly,** go to question 8.
If you answered **incorrectly,** review "Superficial Landmarks" on page 42.

8. What is the name of the first part that is inferior to the posterior brachium?

    Answer: Cubital (olecranon)

If you answered **correctly,** go to question 10.
If you answered **incorrectly,** go to question 9.

9. The brachium is located _____ to the antecubital.

    Answer: superior or proximal

If you answered **correctly,** go to question 10.
If you answered **incorrectly,** review "Superficial Landmarks" on page 42.

10. Posterior to the knee region is the _____.

    Answer: popliteal

If you answered **correctly,** go to question 12.
If you answered **incorrectly,** go to question 11.

11. What is the name of the first part that is superior to the sura?

    Answer: Popliteal

If you answered **correctly,** go to question 12.
If you answered **incorrectly,** review "Superficial Landmarks" on page 42.

12. The area between the right hypochondriac and the left hypochondriac is called the _____.

    Answer: epigastric

If you answered **correctly,** go to question 14.
If you answered **incorrectly,** go to question 13.

13. The ascending colon of the large intestine is located in the _____ region.

    Answer: right lumbar

If you answered **correctly,** go to question 14.
If you answered **incorrectly,** review "Abdominopelvic Regions" on page 51.

14. The plane that passes through the body and separates it into anterior and posterior portions is the _____.

   Answer: frontal

If you answered **correctly,** go to question 16.
If you answered **incorrectly,** go to question 15.

15. The diaphragm muscle actually "divides" the body as a _____ cut.

   Answer: transverse

If you answered **correctly,** go to question 16.
If you answered **incorrectly,** review "Body Planes" on page 56.

16. The body responds to bacteria by creating a fever. This is an example of a _____ feedback mechanism.

   Answer: positive

If you answered **correctly,** go to question 18.
If you answered **incorrectly,** go to question 17.

17. An experiment was done to show that a blood vessel that has a diameter of 2 mm will dilate to 3 mm under certain conditions and the constrict to 1 mm under other conditions. So, over a 20 minute period of time, the blood vessels could dilate and constrict many times depending on the conditions. This is an example of _____.

   Answer: negative feedback

If you answered **correctly,** go to question 18.
If you answered **incorrectly,** review "Negative Feedback Mechanisms" on pages 58–59.

18. According to the periodic table, when calcium becomes an ion it will develop what charge?

   Answer: 2+

If you answered **correctly,** go to question 20.
If you answered **incorrectly,** go to question 19.

19. According to the periodic table, when phosphorus becomes an ion it will develop what charge?

   Answer: 3–

If you answered **correctly,** go to question 20.
If you answered **incorrectly,** review "The Periodic Table" on page 66.

20. What is the chemical symbol for the bicarbonate ion?

    Answer: HCO_3^{1-}

If you answered **correctly,** go to question 22.
If you answered **incorrectly,** go to question 21.

21. What is the chemical symbol of the phosphate ion?

    Answer: PO_4^{3-}

If you answered **correctly,** go to question 22.
If you answered **incorrectly,** review "Ions" on pages 67–69 and "Polyatomic Ions" on pages 69–71.

22. A pH of 3 is how many times more acidic than a pH of 7?

    Answer: 10,000

If you answered **correctly,** go to question 24.
If you answered **incorrectly,** go to question 23.

23. A pH of 8 is how many times more basic than a pH of 6?

    Answer: 100

If you answered **correctly,** go to question 24.
If you answered **incorrectly,** review "pH Concepts" on page 78.

24. Which cell organelle produces protein?

    Answer: Ribosomes

If you answered **correctly,** go to question 26.
If you answered **incorrectly,** go to question 25.

25. The mitochondria produce _____.

    Answer: ATP (adenosine triphosphate)

If you answered **correctly,** go to question 26.
If you answered **incorrectly,** review "The Cell Organelles" on page 83.

26. What is the molecule that "delivers" a message to the ribosomes telling them to make protein?

    Answer: messenger ribonucleic acid (mRNA)

If you answered **correctly,** go to question 28.
If you answered **incorrectly,** go to question 27.

27. Transcription occurs _____ and translation occurs
    _____.

   Answer: at the DNA molecule; at the ribosomes

If you answered **correctly,** go to question 28.
If you answered **incorrectly,** review "Protein Synthesis" on page 83.

28. The most abundant molecule that makes up the cell membrane is the
    _____.

   Answer: phospholipid

If you answered **correctly,** go to question 30.
If you answered **incorrectly,** go to question 29.

29. What is the name of the molecule that will form a channel to allow material to pass in and
    out of the cell?

   Answer: Protein

If you answered **correctly,** go to question 30.
If you answered **incorrectly,** review "The Cell Membrane" on page 84.

30. Describe a hypertonic ECF.

   Answer: The ECF has a higher concentration of solutes than the ICF has.

If you answered **correctly,** go to question 34.
If you answered **incorrectly,** go to question 31.

31. Describe a hypotonic ECF.

   Answer: The ECF has a lower concentration of solutes than the ICF has.

If you answered **correctly,** go to question 34.
If you answered **incorrectly,** review "Osmosis" on pages 85–88.

32. A cell contains 30% water and the environment (ECF) the cell sits in is hypertonic. In order
    for this statement to be true, the percentage of solutes in the ECF has to be
    _____.

   Answer: greater than 70%

If you answered **correctly,** go to question 34.
If you answered **incorrectly,** go to question 33.

33. The environment (ECF) a cell sits in contains 20% solutes and the cell itself is considered to be hypertonic. In order for this statement to be true, the percentage of solutes inside the cell has to be _____.

Answer: greater than 20%

If you answered **correctly,** go to question 34.
If you answered **incorrectly,** review "Osmosis" on pages 85–88.

34. What is the name of the phase of mitosis where the spindle fibers retract and therefore begin to pull the paired chromatids apart?

Answer: Anaphase

If you answered **correctly,** go to question 36.
If you answered **incorrectly,** go to question 35.

35. We can first begin to see the paired chromatids during which phase of cell reproduction?

Answer: Prophase

If you answered **correctly,** go to question 36.
If you answered **incorrectly,** review "Cell Reproduction" on pages 88–89.

36. The main way to identify cardiac cells is to look for
_____.

Answer: intercalated discs

If you answered **correctly,** go to question 38.
If you answered **incorrectly,** go to question 37.

37. The main way to identify cartilage cells is via the presence of
_____.

Answer: a large lacuna

If you answered **correctly,** go to question 38.
If you answered **incorrectly,** review "The Four Tissue Groups" on pages 95–102.

38. Which cell type is classified as having involuntary action other than cardiac cells?

Answer: Smooth muscle cells

If you answered **correctly,** go to question 40.
If you answered **incorrectly,** go to question 39.

39. Which type of cells provides us with our second line of defense?

Answer: Leukocytes (white blood cells)

If you answered **correctly,** go to question 40.
If you answered **incorrectly,** review "The Four Tissue Groups" on pages 95–102.

40. Which integumentary gland is involved in acne production?

> **Answer:** Sebaceous gland

If you answered **correctly,** go to question 42.
If you answered **incorrectly,** go to question 41.

41. Which integumentary gland is involved in cooling the body?

> **Answer:** Merocrine gland

If you answered **correctly,** go to question 42.
If you answered **incorrectly,** review "Glands" on page 111.

42. The mastoid process is _____ (directional term) to the external auditory meatus (external acoustic canal).

> **Answer:** posterior

If you answered **correctly,** go to question 46.
If you answered **incorrectly,** go to question 43.

43. The lateral edge of the sphenoid bone is _____ (directional term) to the temporal bone.

> **Answer:** anterior

If you answered **correctly,** go to question 46.
If you answered **incorrectly,** go to question 44.

44. The lacrimal bone is immediately anterior to the _____ bone.

> **Answer:** ethmoid

If you answered **correctly,** go to question 46.
If you answered **incorrectly,** go to question 45.

45. What is the name of the bone that makes up the posterior one-third of the roof of the mouth?

> **Answer:** Posterior palatine

If you answered **correctly,** go to question 46.
If you answered **incorrectly,** review "Bones of the Face" on pages 117–118.

46. The temporal mandibular joint is made of
_____.

> **Answer:** mandibular fossa of the temporal bone and the mandibular condyle of the mandible

If you answered **correctly,** go to question 48.
If you answered **incorrectly,** go to question 47.

47. The coronoid process of the mandible is _____ (directional term) to the mandibular condyle.

    Answer: anterior

If you answered **correctly,** go to question 48.
If you answered **incorrectly,** review "Bones of the Face" on pages 117–119.

48. From an inferior view, the foramen lacerum is located medial to the foramen _____.

    Answer: ovale

If you answered **correctly,** go to question 49.
If you answered **incorrectly,** go to question 51.

49. From an inferior view, the carotid canal is located _____ (directional term) to the foramen spinosum.

    Answer: posterior

If you answered **correctly,** go to question 50.
If you answered **incorrectly,** review "Foramen of the Skull" on page 120.

50. What is the name of the vertebra that attaches to the skull?

    Answer: Atlas (cervical 1)

If you answered **correctly,** go to question 52.
If you answered **incorrectly,** go to question 51.

51. We have _____ thoracic vertebrae and _____ lumbar vertebrae and _____ cervical vertebrae.

    Answer: 12; 5; 7

If you answered **correctly,** go to question 52.
If you answered **incorrectly,** review "The Vertebral Column" on pages 123–125.

52. The glenoid cavity (fossa) is a/an _____ (directional term) structure on the scapula.

    Answer: lateral

If you answered **correctly,** go to question 56.
If you answered **incorrectly,** go to question 53.

53. The condyle lateral to the trochlea of the humerus is the _____.

    Answer: capitulum

If you answered **correctly,** go to question 56.
If you answered **incorrectly,** go to question 54.

54. Which bone moves when you pronate the antebrachium?

    Answer: Radius

If you answered **correctly,** go to question 56.
If you answered **incorrectly,** go to question 55.

55. Our elbow bone is actually the _____.

    Answer: olecranon

If you answered **correctly,** go to question 56.
If you answered **incorrectly,** review "The Pectoral Girdle and Upper Limbs" on pages 129–131.

56. The head of the femur sits in the _____.

    Answer: acetabulum

If you answered **correctly,** go to question 58.
If you answered **incorrectly,** go to question 57.

57. What is the name of the lateral bone of the lower leg?

    Answer: Fibula

If you answered **correctly,** go to question 58.
If you answered **incorrectly,** review "The Pelvic Girdle and Lower Limbs" on pages 129–131.

58. Describe the location of the extensor digitorum.

    Answer: This muscle is located on the posterior side of the antebrachium.

If you answered **correctly,** go to question 60.
If you answered **incorrectly,** go to question 59.

59. Describe the location of the biceps femoris.

    Answer: This muscle is located on the posterior side of the femur.

If you answered **correctly,** go to question 60.
If you answered **incorrectly,** review "Select Muscles" on pages 141–150.

60. During muscle contraction, the cross-bridges will extend from the _____
    filaments and attach to the _____ filaments.

    Answer: myosin; actin

If you answered **correctly,** go to question 62.
If you answered **incorrectly,** go to question 61.

61.   Which band of the sarcomere consists of overlapping myosin and actin?

      Answer: The "A" band.

If you answered **correctly,** go to question 62.
If you answered **incorrectly,** review "Muscle Structure" on page 150.

62.   Describe a depolarized nerve (neuron).

      Answer: This is a neuron that has a few positive ions (due to a stimulus of some sort) starting to enter the ICF of the neuron.

If you answered **correctly,** go to question 68.
If you answered **incorrectly,** go to question 63.

63.   Describe a nerve (neuron) that is repolarizing.

      Answer: This is a neuron that has started to restore the positive ions from the ICF to the ECF.

If you answered **correctly,** go to question 68.
If you answered **incorrectly,** go to question 64.

64.   What ion causes the release of neurotransmitters?

      Answer: Calcium ions (Ca^{2+})

If you answered **correctly,** go to question 68.
If you answered **incorrectly,** go to question 65.

65.   Neurotransmitters are released from the _____ end of the neuron.

      Answer: axon end (presynaptic vesicle)

If you answered **correctly,** go to question 68.
If you answered **incorrectly,** go to question 66.

66.   A neurotransmitter is decomposed by an enzyme that comes from the _____.

      Answer: dendrite

If you answered **correctly,** go to question 68.
If you answered **incorrectly,** go to question 67.

67.   In order for acetylcholine to be used again, it must be decomposed and then reabsorbed into the _____.

      Answer: presynaptic vesicle

If you answered **correctly,** go to question 68.
If you answered **incorrectly,** review "The Impulse" on pages 156–158.

68. The metencephalon consists of the _____.

   Answer: cerebellum and pons

If you answered **correctly,** go to question 70.
If you answered **incorrectly,** go to question 69.

69. The diencephalon consists of the _____.

   Answer: thalamus and hypothalamus (the pituitary gland can also be included in the diencephalon)

If you answered **correctly,** go to question 70.
If you answered **incorrectly,** review "The Brain" on pages 167–174.

70. Which brain structure "tells" us that we are thirsty from time to time?

   Answer: Hypothalamus

If you answered **correctly,** go to question 72.
If you answered **incorrectly,** go to question 71.

71. Which part of the brain gives us the "urge" to cough from time to time?

   Answer: Medulla oblongata

If you answered **correctly,** go to question 72.
If you answered **incorrectly,** review "The Brain" on pages 167–174.

72. Identify the meninges from the brain tissue to the skull.

   Answer: Pia mater, subarachnoid, dura mater

If you answered **correctly,** go to question 74.
If you answered **incorrectly,** go to question 73.

73. The cerebrospinal fluid is produced by the _____ and serves to help protect the brain and spinal cord.

   Answer: choroid plexus

If you answered **correctly,** go to question 74.
If you answered **incorrectly,** review "Protecting the Central Nervous System" on page 179.

74. Which of the autonomic nerves is involved in pupil dilation?

   Answer: Sympathetic

If you answered **correctly,** go to question 76.
If you answered **incorrectly,** go to question 75.

75.   Which of the autonomic nerves decreases the heart rate?

      Answer: Parasympathetic

If you answered **correctly,** go to question 76.
If you answered **incorrectly,** review "The Spinal Nerves" on page 186.

76.   The trochlear nerve is cranial nerve number _____.

      Answer: IV

If you answered **correctly,** go to question 78.
If you answered **incorrectly,** go to question 77.

77.   The vagus nerve is cranial nerve number _____.

      Answer: X

If you answered **correctly,** go to question 78.
If you answered **incorrectly,** review "The Cranial Nerves" on pages 187–188.

78.   The specialized cells associated with the retina of the eye are called
      _____ and _____.

      Answer: rods; cones

If you answered **correctly,** go to question 80.
If you answered **incorrectly,** go to question 79.

79.   The cells that make up the taste buds are called _____ cells.

      Answer: gustatory

If you answered **correctly,** go to question 80.
If you answered **incorrectly,** review "Introducing the Five Senses" on pages 195–196.

80.   The neurohypophysis releases _____ and
      _____.

      Answer: oxytocin and antidiuretic hormone

If you answered **correctly,** go to question 82.
If you answered **incorrectly,** go to question 81.

81.   What hormone comes from the adenohypophysis and targets the adrenal cortex?

      Answer: Adrenocorticotropic Hormone (ACTH)

If you answered **correctly,** go to question 82.
If you answered **incorrectly,** review "The Pituitary Gland" on pages 207–209.

82. The thyroid gland releases _____.

    Answer: calcitonin or thyroxine or triiodothyronine

If you answered **correctly,** go to question 86.
If you answered **incorrectly,** go to question 83.

83. Adrenalin is produced by what gland?

    Answer: Adrenal medulla

If you answered **correctly,** go to question 86.
If you answered **incorrectly,** go to question 84.

84. Erythropoietin will target the _____ and cause
    _____.

    Answer: bone marrow; red blood cell formation

If you answered **correctly,** go to question 86.
If you answered **incorrectly,** go to question 85.

85. The thyroid stimulating hormone comes from the _____ and
    targets the _____.

    Answer: adenohypophysis; thyroid gland

If you answered **correctly,** go to question 86.
If you answered **incorrectly,** review "Other Endocrine Glands" on pages 209–211.

86. Identify the hormone that causes the nephrons to lose sodium ions to the toilet?

    Answer: Atrial natriuretic peptide (ANP)

If you answered **correctly,** go to question 90.
If you answered **incorrectly,** go to question 87.

87. If a patient had very little sodium ions in their blood, they would produce excess amounts
    of _____ to bring the sodium ion levels back to normal.

    Answer: aldosterone

If you answered **correctly,** go to question 90.
If you answered **incorrectly,** go to question 88.

88. A lack of _____ will cause excess glucose in the urine.

    Answer: insulin

If you answered **correctly,** go to question 90.
If you answered **incorrectly,** go to question 89.

89. A lack of _____ could result in a decrease in blood calcium levels.

    Answer: parathormone

If you answered **correctly,** go to question 90.
If you answered **incorrectly,** review "Other Endocrine Glands" on pages 209–211.

90. What causes the release of EPO?

    Answer: A decrease in oxygen going to the kidney cells.

If you answered **correctly,** go to question 94.
If you answered **incorrectly,** go to question 91.

91. EPO targets _____.

    Answer: bone marrow

If you answered **correctly,** go to question 94.
If you answered **incorrectly,** go to question 92.

92. Erythrocytes have a lifespan of only 120 days. They do not live very long because they have neither _____.

    Answer: a nucleus nor any cell organelles

If you answered **correctly,** go to question 94.
If you answered **incorrectly,** go to question 93.

93. What is the main protein molecule found inside an erythrocyte?

    Answer: Hemoglobin

If you answered **correctly,** go to question 94.
If you answered **incorrectly,** review "Red Blood Cells" on pages 218–219.

94. What is a normal percent value for neutrophils in a non-sick patient?

    Answer: 40% to 60%

If you answered **correctly,** go to question 96.
If you answered **incorrectly,** go to question 95.

95. What is a normal percent value for eosinophils in a non-sick patient?

    Answer: 3%

If you answered **correctly,** go to question 96.
If you answered **incorrectly,** review "White Blood Cells" on pages 219–220.

96. What is the name for blood clotting factor IV?

    Answer: Calcium ions (Ca^{2+})

If you answered **correctly,** go to question 98.
If you answered **incorrectly,** go to question 97.

97. What is the name for blood clotting factor VIII?

    Answer: Antihemophiliac factor

If you answered **correctly,** go to question 98.
If you answered **incorrectly,** review "Platelets and Platelet Response" on pages 221–222.

98. People with blood type A have agglutinogen _____ on their erythrocytes and agglutinin _____ in their plasma.

    Answer: A; b

If you answered **correctly,** go to question 100.
If you answered **incorrectly,** go to question 99.

99. People with blood type B have agglutinogen _____ on their erythrocytes and agglutinin _____ in their plasma.

    Answer: B; a

If you answered **correctly,** go to question 100.
If you answered **incorrectly,** review "Glycolipids and Blood Typing" on pages 224–228.

100. Is the following packed cell donation safe? Donate type AB to type B?

    Answer: No

If you answered **correctly,** go to question 104.
If you answered **incorrectly,** go to question 101.

101. Is the following packed cell donation safe? Donate type B to type AB?

    Answer: Yes

If you answered **correctly,** go to question 104.
If you answered **incorrectly,** go to question 102.

102. Is the following whole blood donation safe? Type B to type AB?

    Answer: No

If you answered **correctly,** go to question 104.
If you answered **incorrectly,** go to question 103.

103. Is the following plasma donation safe? Type AB to type B?

    Answer: Yes

If you answered **correctly,** go to question 104.
If you answered **incorrectly,** review "Donating Packed Cells," "Donating Whole Blood," and "Donating Plasma" on pages 224, 228, and 230.

104. People with blood type A– have what kind of agglutinogen/s and agglutinin/s?

    Answer: A agglutinogens and b agglutinins

If you answered **correctly,** go to question 106.
If you answered **incorrectly,** go to question 105.

105. People with blood type AB+ have what kind of agglutinogen/s and agglutinin/s?

    Answer: A agglutinogens and B agglutinogens and D agglutinogens and no agglutinins.

If you answered **correctly,** go to question 106.
If you answered **incorrectly,** review "The Rh Factor" on pages 233–235.

106. In order for blood to enter into the right ventricle, the blood must pass through the
    _____ valve.

    Answer: tricuspid

If you answered **correctly,** go to question 107.
If you answered **incorrectly,** go to question 117.

107. Blood in the pulmonary trunk came from which chamber of the heart?

    Answer: Right ventricle

If you answered **correctly,** go to question 108.
If you answered **incorrectly,** review "Structures of the Internal Heart" on pages 239–242.

108. What does the p wave of an ECG recording represent?

    Answer: The p wave represents atrial depolarization, which is the travel of the impulse from the SA node to the AV node.

If you answered **correctly,** go to question 110.
If you answered **incorrectly,** go to question 109.

109. The largest bump on an ECG is represented by the letters _____ and is the recording of which heart activity?

    Answer: QRS; the QRS represents the depolarization of the ventricles, which is a recording of the nerve activity within the Purkinje fibers. The nerves of the atria are also repolarizing at this time.

If you answered **correctly,** go to question 110.
If you answered **incorrectly,** review "The Electrocardiogram (ECG)" on pages 245–246.

110. Blood in the right subclavian artery will enter into the _____ artery next.

    Answer: axillary

If you answered **correctly,** go to question 112.
If you answered **incorrectly,** go to question 111.

111. Blood from the brachial artery will flow into the _____ artery next.

    Answer: radial and ulnar

If you answered **correctly,** go to question 112.
If you answered **incorrectly,** review "Blood Vessels (Arteries)" on pages 247–253.

112. Blood in the popliteal vein will enter into the _____ vein next.

    Answer: femoral

If you answered **correctly,** go to question 114.
If you answered **incorrectly,** go to question 113.

113. Blood in the right subclavian vein will enter the _____ vein next.

    Answer: right brachiocephalic

If you answered **correctly,** go to question 114.
If you answered **incorrectly,** review "Blood Vessels (Veins)" on pages 254–261.

114. The glycolipid associated with erythrocytes is called a/an _____ and the glycolipid associated with disease causing organisms is called a/an _____.

    Answer: agglutinogen; antigen

If you answered **correctly,** go to question 118.
If you answered **incorrectly,** go to question 115.

115. What type of lymphocytes are responsible for producing antibodies?

    Answer: B cells

If you answered **correctly,** go to question 118.
If you answered **incorrectly,** go to question 116.

116. When we come in contact with a viral antigen, we develop immunity against that specific antigen. However, that same viral antigen may attack us again next year. We are immune to it because our _____ will respond immediately.

    Answer: memory B cells

If you answered **correctly,** go to question 118.
If you answered **incorrectly,** go to question 117.

117. The reason we become sick with the flu a second time (and more) is because the second time was caused by a flu virus that had a different _____.

   Answer: antigen

If you answered **correctly,** go to question 118.
If you answered **incorrectly,** review "The Lymphatic System and Defense" on pages 265–270.

118. The carina is the point where the _____.

   Answer: trachea branches to form the two primary bronchi

If you answered **correctly,** go to question 120.
If you answered **incorrectly,** go to question 119.

119. Which respiratory tubes do not consist of any cartilage?

   Answer: The bronchioles lack cartilage.

If you answered **correctly,** go to question 120.
If you answered **incorrectly,** review "The Respiratory Organs" on page 273.

120. If the _____ tonsil swells it would make breathing through the nose very difficult.

   Answer: pharyngeal

If you answered **correctly,** go to question 122.
If you answered **incorrectly,** go to question 121.

121. The _____ tonsils are located in the nasopharynx region.

   Answer: pharyngeal

If you answered **correctly,** go to question 122.
If you answered **incorrectly,** review "The Respiratory Organs" on page 274.

122. In order to inhale, the diaphragm muscle must move in which direction?

   Answer: Downward (inferiorly)

If you answered **correctly,** go to question 124.
If you answered **incorrectly,** go to question 123.

123. When the diaphragm muscle is contracting, it is moving in which direction?

   Answer: Downward

If you answered **correctly,** go to question 124.
If you answered **incorrectly,** review "The Process of Inhaling and Exhaling" on page 275.

124. If you increase the air pressure inside your thoracic cavity, air will
_____.

   Answer: go out of the lungs

If you answered **correctly,** go to question 126.
If you answered **incorrectly,** go to question 125.

125. If the outside air pressure is 780 mm Hg, air will enter the lungs if we
_____ the size of our thoracic cavity to change the air pressure
to _____ mm Hg.

   Answer: increase; a value less than 780 mm Hg.

If you answered **correctly,** go to question 126.
If you answered **incorrectly,** review "The Process of Inhaling and Exhaling" on page 275.

126. What percentage of the carbon dioxide we produce will eventually form hydrogen ions in
the red blood cell?

   Answer: 70%

If you answered **correctly,** go to question 128.
If you answered **incorrectly,** go to question 127.

127. What percentage of the carbon dioxide generated is normally exhaled?

   Answer: 23%

If you answered **correctly,** go to question 128.
If you answered **incorrectly,** review "The Chloride Shift" on pages 277–279.

128. If a person is already in the process of hyperventilating, their blood pH will
_____.

   Answer: go up

If you answered **correctly,** go to question 130.
If you answered **incorrectly,** go to question 129.

129. Carbon dioxide binds with water and thus produces _____,
which ionizes and forms _____ and _____.

   Answer: carbonic acid; hydrogen ions and bicarbonate ions

If you answered **correctly,** go to question 130.
If you answered **incorrectly,** review "Abnormal Breathing" on pages 281–282.

130. Nutrients will pass from the small intestine into the bloodstream by being absorbed through the _____.

    Answer: villi

If you answered **correctly,** go to question 132.
If you answered **incorrectly,** go to question 131.

131. The pancreas produces hormones and _____.

    Answer: digestive enzymes

If you answered **correctly,** go to question 132.
If you answered **incorrectly,** review "The Small Intestine" on pages 292–293 and "The Pancreas" on pages 296–297.

132. Identify the digestive enzymes that come from the pancreas and digests protein.

    Answer: Trypsin, chymotrypsin, and carboxypeptidase

If you answered **correctly,** go to question 137.
If you answered **incorrectly,** go to question 133.

133. Identify the digestive enzyme that comes from the pancreas and digests fat.

    Answer: Lipase

If you answered **correctly,** go to question 137.
If you answered **incorrectly,** go to question 134.

134. Identify an enzyme that is produced by the small intestine and digests protein.

    Answer: Peptidase

If you answered **correctly,** go to question 137.
If you answered **incorrectly,** go to question 135.

135. Identify an enzyme that comes from the stomach and digest material to form amino acids.

    Answer: Pepsin

If you answered **correctly,** go to question 137.
If you answered **incorrectly,** go to question 136.

136. Identify an enzyme that comes from the pancreas and digests material to form fatty acids.

    Answer: Lipase

If you answered **correctly,** go to question 137.
If you answered **incorrectly,** review "Summary of Digestive Enzymes" on page 302.

137. Identify the hormone that will cause the production of bile.

   Answer: Secretin

If you answered **correctly,** go to question 142.
If you answered **incorrectly,** go to question 138.

138. Identify the hormone that causes the gallbladder to release bile.

   Answer: Cholecystokinin (CCK)

If you answered **correctly,** go to question 142.
If you answered **incorrectly,** go to question 139.

139. Identify the hormone that is responsible for helping the small intestine maintain an optimum pH value.

   Answer: Secretin

If you answered **correctly,** go to question 142.
If you answered **incorrectly,** go to question 140.

140. Identify the hormone that causes the pancreas to release buffers.

   Answer: Secretin

If you answered **correctly,** go to question 142.
If you answered **incorrectly,** go to question 141.

141. Identify the hormone that causes the production of stomach acids.

   Answer: Gastrin

If you answered **correctly,** go to question 142.
If you answered **incorrectly,** review "Hormones of the Digestive System" on page 303.

142. Glycolysis is a series of chemical reactions that occur in the _____ and the Krebs' reactions are a series of chemical reactions that occur in the _____.

   Answer: cytosol of the cell; mitochondria

If you answered **correctly,** go to question 145.
If you answered **incorrectly,** go to question 143

143. The conversion of glucose ultimately to pyruvic acid is known as _____.

   Answer: glycolysis

If you answered **correctly,** go to question 145.
If you answered **incorrectly,** go to question 144

144. Basically, all the food we eat will become nutrients and will go to what cell organelle?

   Answer: Mitochondria

If you answered **correctly,** go to question 145.
If you answered **incorrectly,** review "Metabolism of Carbohydrates" on pages 308–309.

145. What are essential amino acids?

   Answer: Essential amino acids are the amino acids that we have to obtain from a good diet. The body cannot make these amino acids.

If you answered **correctly,** go to question 147.
If you answered **incorrectly,** go to question 146.

146. What are nonessential amino acids?

   Answer: Nonessential amino acids are the amino acids that the body can produce. It is NOT essential to get these from the diet.

If you answered **correctly,** go to question 147.
If you answered **incorrectly,** review "Essential and Nonessential Products" on pages 313–314.

147. Some of the excess cholesterol that our body encounters will be transported to the liver by _____. In the liver, this cholesterol will be incorporated into _____.

   Answer: high density lipoprotein (HDL); bile

If you answered **correctly,** go to question 149.
If you answered **incorrectly,** go to question 148.

148. Why is "good cholesterol" called good?

   Answer: "Good cholesterol" is only good because it is the cholesterol that is ultimately transported out of the body.

If you answered **correctly,** go to question 149.
If you answered **incorrectly,** review "Cholesterol and Metabolism" on page 317.

149. The _____ exit the kidneys and the _____ exits the urinary bladder.

   Answer: ureters; urethra

If you answered **correctly,** go to question 151.
If you answered **incorrectly,** go to question 150.

150. Which kidney sits higher in the body than the other kidney?

   Answer: The left is higher than the right.

If you answered **correctly,** go to question 151.
If you answered **incorrectly,** review "The Functions and Structures of the Urinary System" on page 321.

151. What are the major structures that pass through the renal pyramids?

    Answer: Collecting tubules

If you answered **correctly,** go to question 153.
If you answered **incorrectly,** go to question 152.

152. Water and/or waste that are in the collecting tubules will enter the
    _____ next.

    Answer: minor calyx (minor calyces)

If you answered **correctly,** go to question 153.
If you answered **incorrectly,** review "The Internal Structures of the Kidney" on pages 323–324.

153. When we say that the kidneys are putting water back into the blood, they actually are
    putting water back into the _____.

    Answer: vasa recta

If you answered **correctly,** go to question 155.
If you answered **incorrectly,** go to question 154.

154. Waste products will enter into the nephron after being "forced" out of the
    _____.

    Answer: glomerular capillaries

If you answered **correctly,** go to question 155.
If you answered **incorrectly,** review "Blood Vessels Associated with the Nephrons" on page 326.

155. The action of ADH will put approximately _____% of the water back into the
    bloodstream.

    Answer: 99%

If you answered **correctly,** go to question 157.
If you answered **incorrectly,** go to question 156.

156. The main functioning unit of the kidneys is the _____.

    Answer: nephron

If you answered **correctly,** go to question 157.
If you answered **incorrectly,** review "Function of the Nephron" on page 327.

157. The sperm cells are produced in special tubes inside the testes called
    _____.

    Answer: seminiferous tubules

If you answered **correctly,** go to question 159.
If you answered **incorrectly,** go to question 158.

158. The tube that transports the sperm from the testes to the penile urethra is called
_____.

Answer: ductus deferens

If you answered **correctly,** go to question 159.
If you answered **incorrectly,** review "The Male Reproductive System and Sperm Cells" on page 335.

159. What hormone causes the corpus luteum to continue producing hormones beyond the normal 10 day time period?

Answer: Human chorionic gonadotropin (HCG)

If you answered **correctly,** you are finished! Congratulations.
If you answered **incorrectly,** go to question 160.

160. A decrease in what hormone will initiate menstruation?

Answer: Progesterone

If you answered **correctly,** you are finished! Congratulations.
If you answered **incorrectly,** go to question 161.

161. A ruptured follicle is called a _____.

Answer: corpus luteum

If you answered **correctly,** you are finished! Congratulations.
If you answered **incorrectly,** go to question 162.

162. What hormone/s does the ruptured follicle produce?

Answer: Progesterone and estrogen

If you answered **correctly,** you are finished! Congratulations.
If you answered **incorrectly,** go to question 163.

163. A decrease in what hormone will initiate menopause?

Answer: Follicle stimulating hormone

If you answered **correctly,** you are finished! Congratulations.
If you answered **incorrectly,** go to question 164.

164. The presence of what hormone indicates pregnancy?

Answer: Human Chorionic Gonadotropin (hCG)

If you answered **correctly,** you are finished! Congratulations.
If you answered **incorrectly,** review "Hormones Associated with the Female Reproductive System" on page 341.

Index

A

abdomen, 42
abdominal region, 49
abdominopelvic regions
    appendix, 53
    ascending colon, 53
    cecum, 53
    defined, 49, 51
    descending colon, 53
    epigastric, 51
    example problems, 52–53, 54
    hypogastric, 51
    illustrated, 52
    left hypochondriac, 51
    left inguinal, 51
    left lumbar, 51
    liver, 53
    physician use, 53
    right hypochondriac, 51
    right inguinal, 51
    right lumbar, 51
    small intestine, 53
    spleen, 53
    stomach, 53
    umbilical, 51
    urinary bladder, 53
    work problems, 54–55
    worked solutions, 55–56
abducens nerve, 187
abnormal breathing, 281–282
acetylcholine
    breakdown, 158
    defined, 158
    release, 157
acetylcholinesterase, 157
acid reflux, 291
acid-base balance, 331
acidic chyme, 292
acrosomal cap, 337
actin
    binding sites, 151
    cross-bridge binding, 151
    defined, 150
    movement, 150
    myofilament, 151
action potential, 156
active immunity, 268
adenosine triphosphate. *See* ATP
adipose cells, 99
adrenal cortex, 210
adrenal medulla, 210
adrenocorticotropic hormone (ACTH)
    function, 208
    location, 207
    target, 208
afferent arteriole, 326

afferent nerves, 185
agglutination, 226, 235
agglutinins
    defined, 224
    in packed cell donation, 226, 227
    in plasma donation, 230, 231
    in Rh factor donations, 234
    in whole blood donation, 228, 229
agglutinogen, 224
aldosterone
    defined, 332
    function, 211
    location, 210
    target, 210
alveolar macrophages, 283
alveoli, 273, 274
amino acids. *See also* proteins
    bonding sequences, 309
    break down, 310
    in catabolic reactions, 308
    defined, 77
    essential, 313–314
    formation, 307
    hemoglobin, 218
    nonessential, 313–314
    number of, 77, 309
    sequences, 309
ammonium ions, 69, 70, 71
anabolism. *See also* metabolism
    activities, 308
    carbohydrates, 309
    defined, 307, 308
    example problems, 308
    fats, 311
    proteins, 310
anaphase, 89
anatomical position, 38
ankle
    components, 134
    cuboid bone, 135
    cuneiform bones, 134, 135
    navicular, 134, 135
    talus, 134, 135
ANP. *See* atrial natriuretic peptide
antebrachium, 42
antecubital, 42
anterior, 38
anterior gray, 178
anterior palatine, 118
anterior skeleton, 135
anterior white, 178
antibodies, 267, 268
antidiuretic hormone (ADH)
    function and, 208
    location, 207
    release, 332
    target, 208
    water and, 328

CPSIA information can be obtained at www.ICGtesting.com
Printed in the USA
LVOW09s1854231013

358275LV00001BA/93/P

9 780764 574696